EG Council Tax Handbook

Geoff Parsons
with
Tim Rowcliffe Smith

2006

Routledge
Taylor & Francis Group

LONDON AND NEW YORK

First published 2006 by Estates Gazette

Published 2014 by Routledge
2 Park Square, Milton Park, Abingdon, Oxon OX14 4RN
711 Third Avenue, New York, NY 10017, USA

Routledge is an imprint of Taylor & Francis Group, an informa business

ISBN 13 : 978-0-7282-0484-3 (pbk)

Typeset in Palatino 10/12 by Amy Boyle, Rochester

Contents

Part 1 Council Tax and its History

Part 2 Organisations and Structures

Part 3 Property and Management

Part 4 Administration and Recovery of Council Tax

Part 5 Council Tax Payers

Part 6 Valuation Lists and Assessment

Part 7 Dispute Resolution

Part 8 Additions or Alternatives to Council Tax

Part 9 Appendices

Part 10 Indexes

Acknowledgments

Local taxation has a very long history in the United Kingdom and has been explored generally and specifically by writers, reviewers, historians, committees of inquiry, judges and tribunal members, officials and others. The list of readings indicates a few of the texts which readers may find particularly helpful — we have perused many of them and wish to acknowledge with our appreciation those who worked on this material, particularly those responsible for the many documents published by the Audit Commission, the Institute of Revenues, Rating and Valuation, the Valuation Office Agency, and the Valuation Tribunal Service. Their writings gave ideas and insights which have proved invaluable in developing the many themes in this volume.

We acknowledge the personal help of several individuals who have kindly read some of our draft material and commented on it. In most instances we were able to make changes and incorporate or develop the thoughts given. Our appreciation of such help from the following is substantial: Brian Daley, Daniel Hilton, Councillor Frank Parker of Sevenoaks District Council, Sharon Smartt and finally but not least Mayor Barrie Wigg of Swanley Council. Nevertheless, any omissions and errors must be acknowledged as our own. We thank too members of the staff of the IRRV, the Kent Library Service and both the Citizens Advice Bureau and Volunteer Bureau at Swanley who assisted in many different ways. Also, our thanks go to Louise Newton for assistance with the indexes.

At the Estates Gazette advice and guidance was readily given by Audrey Andersson, Amy Boyle, Rebecca Chakaborty and Alison Richards — they answered our questions and ensured we were made aware of matters of publishing importance — nevertheless any failings must be put down to the authors!

Finally, we have both received, with much appreciation, the support of our families which proved crucial to our capacity to "get on with" the research and writing needed to complete the text.

Preface

With recent revaluations in Wales and Northern Ireland and one in England due in the future, the *EG Council Tax Handbook* seeks to address the needs of those who want an easy to read but comprehensive text on the subject. In particular it addresses the following:

- the detail of council tax as it is administered and enforced
- the range of exemptions, discounts, deductions and reliefs available, including council tax benefit
- the history of local taxation so as to explore the nature of or define the principles which underlie local taxes
- the compilation and maintenance of the valuation lists
- the basis of and practical points on valuation
- the approaches to dispute resolution and their application to council tax and council tax benefit
- the search for possible additions or alternatives to council tax.

The book should meet the needs of owners, occupiers and their professional advisers who face increased assessments as a result of the revaluations. That is, it should meet the needs of those who need insights, for instance, into one or more of the following:

- the dwellings which are covered or are exempt
- the individuals who must pay, or otherwise may enjoy an exemption or be entitled to a relief from the tax, eg as a result of the transitional arrangement following revaluation
- what will happen if a dwelling is newly created, altered, partly demolished or used for non-domestic purposes

Of course, disputes arise in many council tax and council tax benefit situations, including those:

- between the listing officer and the owner or occupier regarding an assessment
- between a billing authority and a taxpayer or a claimant of council tax benefit
- between a landlord and his or her lessee or tenant.

This volume should give step by step insights into the procedures for such bodies as the valuation tribunal, the magistrate's court and the ombudsman. Also, where a billing authority or other organisation has alleged that a taxpayer or some other person has not paid council tax, this book outlines such matters as enforcement practice, ie liability orders — leading on to the likes of bailiff's action, attachment orders and even committal to prison. The text seeks to cover counter-fraud in some detail by outlining the organisation and structure, the offences which occur and the administrative or prosecution counter-fraud measures which the authorities adopt.

The future of local taxation is considered in the last chapter. In recent years several important thrusts have appeared which have, will or may result in changes to local domestic taxation in the United Kingdom. They are:

- a revaluation for council tax has place in Wales
- a revaluation for council tax in England — to be recommenced in the future
- a revaluation for domestic rating has been completed in Northern Ireland
- local taxation has been reviewed in Northern Ireland and Wales
- reviews of local taxation are underway in Scotland and England.

At the same time, growing taxpayers' dissatisfaction with council tax (as expressed by many individuals and by pressure groups) has resulted in recent pre-election promises which included the replacement of council tax with a local income tax, the abolition of the revaluation in England, a rebate £200 of the council tax for pensioners and prospective changes in the light of the enquiry being conducted by Sir Michael Lyons, eg increasing the number of bands. The Association of Local Authorities has already sounded a request for statutory intervention — for that is what it will mean — to widen the scope of local taxation. This volume adopts the perspective that it is possible to have a wider tax base which could include one or more of a local sales tax, local income tax, site value rating and a host of user-pays charges.

The book contains insights into the workings of dozens of organisations, offices, officials and others who are involved in local taxation policy-making, property assessment, administration, enforcement, counter-fraud investigation and prosecution, dispute determination and the measurement of performance and best value. Of course, the book does not hold all the answers. It gives insights but will not obviate the professional administrator, property professional, the academic, student or the taxpayer who is not professionally grounded in council tax matters from seeking appropriate professional advice. Those who need advice or are not working on a day to day basis in local taxation may seek the following from this volume's 19 chapters:

- understanding of the terms or jargon used by professional advisers
- understanding of the main principles and practices of council tax and council tax benefit law
- essential references to the provisions of Acts of parliament and other enactments as well as reports of court and tribunal decisions
- identification of the issues surrounding their problems in those terms
- knowledge of what information, documents and opinions are relevant to their problem
- appreciation that there may be more than one route to a solution, eg a complaint may be taken in several ways
- understanding the steps to be followed in many procedures
- appreciation of the consequences of action (or in-action)
- understanding of counter-fraud and the performance regimes now in place in local government.

Although the book will appeal to those directly involved with council tax, such as administrators, assessors, regulator, adjudicators and students of cognate disciplines, it should also be of value to those, including those overseas who need an overview — both historic and futuristic — of the way in which local taxation has developed and might develop in the future.

Authors' Note

This volume, *EG Council Tax Handbook*, provides many insights into council tax, council tax benefit and, conceivable prospective changes to local taxation in the United Kingdom. It has 10 Parts — the first eight comprise 19 chapters.

1 *Council tax and its history*: in this Part the two chapters outline the nature of council tax and examine the history of local taxation for principles.
2 *Organisations and structures*: the three chapters look at government, professional and other bodies concerned with council tax. It also covers the sources and uses of information on local taxation.
3 *Property and management*: the two chapters consider the domestic properties covered by council tax and the main property management and works issues arising in respect of the tax.
4 *Administration and recovery of council tax*: two of the four chapters examine office organisation and the duties performed in administering council tax, including demand and collection, enforcement or recovery and council tax benefit. The other chapters consider counter-fraud measures and best value in the context of the measurement of performance.
5 *Council tax payers*: the two chapters seek identify council tax payers and those who enjoy exemptions, discounts, other reductions and reliefs.
6 *Valuation lists and assessment*: in this Part the three chapters consider the procedures of revaluation and the basis of valuation or assessment of dwellings and some practical valuation points.
7 *Dispute resolution*: two chapters outline the approaches to resolving disputes and the roles and activities of those involved in the settlement of disputes.
8 *Additions or alternatives to council tax*: the single chapter of the final Part reviews various possible additions or alternatives to council tax — a tax which is increasingly causing concern.

In addition, it is hoped that all readers will find particular value in the chapters' illustrative boxes and figures of which there are over 70. They may be categorised to cover the following:

- guides to professionals and what they do
- step by step guides to procedures
- profiles of "organisations" or other aspects of local taxation
- information in concise but comprehensive text.

Parts 9 and 10 are appendices and indexes — the readers' "access mechanisms" to the 19 chapters.

They include a list of the boxes; lists of enactments, readings, abbreviations and cases; and, two indexes — organisations and key words. Many enactments are referred to in this volume and specific note of amendments to some of them by later provisions have been included. However, any reader who wishes to trace a particular amendment pathway will need to refer elsewhere since it has not been appropriate to give every amending citation.

The systems of council tax and council tax benefit are very similar in the three countries of Great Britain but differ in detail, eg the council tax bands are not the same. (Northern Ireland's local tax is domestic rating.) Broadly, when it was formulated the legislation for council tax as applied to England and Wales was adapted for Scotland's law and administrative system. Thus, although English law and terminology has been used throughout the book, the reader may like to be mindful when reading the text that Scotland has its own law and administrative structure governing the application of council tax and council tax benefit there.

At the beginning of the book, the contents pages show each chapter's main subsections. Each chapter begins with an "aim", ie the main reason for the chapter. The "objectives" follow as targets for the reader. As mentioned above, the text of each chapter is illustrated with boxes, providing a considerable, but concise, insight to council tax and council tax benefit within the overall context of changing local taxation.

Terms used in this volume follow the relevant meaning given in the text, a cited enactment or that given in the second edition of *The Glossary of Property Terms* which is published by EG Books. (Alternatively, a term will usually be found on the publisher's website *www.egi.com*.)

Geoff Parsons
Tim Rowcliffe Smith
December 2005

Part 1

Council Tax and its History

Council Tax — An Overview

1

Aim

To briefly introduce council tax in Great Britain and domestic rating in Northern Ireland

Objectives

- **to briefly examine the nature of the government's role and activities in local taxation**
- **to describe the role of the listing officer's function of assessment**
- **to outline the way in which council tax is administered by billing authorities and others**
- **to describe the roles of the courts, tribunals and ombudsmen**
- **to briefly consider domestic rating in Northern Ireland**

Introduction

Local taxation has been developed during the 400 years or so since the coming into force of the Poor Relief Act 1601 (the Statute of Elizabeth), itself a consolidation Act. For domestic property, it settled down as rating for most of the 20th century, based upon the rental value of the occupier's residence. However, in the late 1980s rating was replaced by the community charge, which was itself quickly abandoned for council tax. Thus, a government consultation paper of 23 April 1991, *A New Tax for Local Government*, proposed a new tax to take the place of the community charge. As a result, council tax was introduced under the Local Government Finance Act 1992 and began to operate in Great Britain from 1 April 1993.

In broad terms council tax is paid to local councils by local people, mainly occupiers of dwellings, for the services that local authorities and other bodies provide.

The content of this chapter is intended as a framework for the more detailed treatment in later chapters. The chapter begins with a brief review of the legal framework, governmental roles and the general financial context for council tax. It concludes with a brief consideration of domestic rating in Northern Ireland.

Legal framework

Council tax is derived from Acts of parliament and statutory instruments — as interpreted in case law. Frequent references are made to the enactments and cases to identify the authority for various points of law and to direct the interested reader who wishes to go into greater detail. Council tax is in many ways similar to its predecessors — poor relief, domestic and non-domestic rating and community charge. Thus, the council tax enactments are sometimes interpreted by reference to case law that has emerged over the last 400 years or so. For example, *Sir Anthony Earby's Case* (1633) determined that a person can only be assessed on assets owned within the parish, and in distress *Semayne's Case* (1604) (1558–1774) All ER Rep 62, (1604) 5 Co Rep, a limited forcible entry by bailiffs — on the finding that "... the house of everyone is to him as his castle and fortress", ie an "Englishman's castle".

Governmental functions

Central and devolved government

Central government has a relatively discrete set of functions which directly or indirectly affect the funding of local government. Devolved governments in the United Kingdom have similar functions within the framework established by central government. They are briefly described in Box 1.1.

Box 1.1 Functions of government in local taxation

Policy development	• developing policies for local taxation in the context of the socio-economic development of the economy
Law-making	• creating enactments for local taxation and for adopted central policies which affect local government funding directly or indirectly
Taxing	• raising funds by devising and operating a tax base which meets needs and accords with broadly acceptable principles of taxation (see Chapter 2)
Funding	• allocating funds to local government for their duties and discretionary work, particular programmes and projects that are required to support central policies

---

Local government functions

Local authorities have duties which involve delivering policies in the form of services, programmes and projects that have been established at national, devolved or regional levels — sometimes under policies of the European Union. At the same time they have discretionary powers to effect local policies.

As "billing authorities", some of the local authorities, eg district councils, administer the council tax and council tax benefit regimes. They are defined by the Local Government Finance Act 1992 (the 1992 Act) and other enactments. The administration of council tax and council tax benefit is dealt with in detail in Part 4 of this volume.

Financial context for local taxation

This volume is being written at a time when a discussion on the future of local government funding is underway. At present it seems that the government does not intend to make substantial changes to council tax in England. Nevertheless, changes to local taxation are likely in the context of:

- a search for a 10-year strategy on the future of local government
- the balance of funding issue
- the need for additional sources for the local tax base
- the imposing of further "central" services and projects on local government.

Funding of local government is obtained locally and centrally in proportions of around 25% to 75% respectively. About £20 billion in council tax is collected by the local government authorities, ie they are called "billing authorities" in England and Wales and "levying authorities" in Scotland. The latter term is also used for fire services in Wales instead of "precepting authorities". (In this volume they are generally referred to as "billing authorities' unless the context is Scotland.) The money is for direct expenditure by the billing and other local bodies called "precepting authorities" (see section 39(1) and (2) of the 1992 Act).

Generally, the local bodies are free to fix the level of taxation, but there is a ministerial power to cap the amount any particular authority intends to charge (see p12). Each year additional funds, amounting to about £19 billion, come from the non-domestic (business) rates which the local charging bodies had previously remitted to the government — the amounts are redistributed on a capita basis according to prescribed formulae.

Council tax is but one of several sources of money for local services and administration. Box 1.2 lists the main sources and gives a brief description of each.

Proportions of local to central funding

Several issues arise from the balance of funding in terms of local and central sources. Two linked issues are:

- the accountability issue
- the "gearing" problem.

Concerns are expressed about the way in which central funding at roughly the 75% level is said to take accountability or decision-making away from the local level. Similarly, with the level of central funding at about 75%, an increase of one percent in local expenditure requires a four percent increase in council tax (if the central funding remains unchanged).

Box 1.2 Principal sources of local government finance

National non-domestic rating	• rates on non-domestic property • collected by local charging authorities and remitted to the government • the government redistributes the rates to local authorities
Council tax	• tax on domestic and other residential accommodation • collected by billing authorities • some passed to the precepting authorities • many exemptions and reliefs (see Chapter 13)
Receipts from trading, tolls and receipts for services	• planning fees • building control fees • licences fees • concessionaires
Garden square rates	• rates for the upkeep of London squares • collected by some London borough councils
Water charges rates	• rates collected in Scotland • payment for certain water supplies
Municipal bonds	• issues of bonds for local capital projects (not revenue expenditure)
Rents	• rents from housing and non-domestic lettings • fees from licences to use accommodation
Asset sales	• funds from the sale of land and other assets • limitations on availability and local use of these funds
Government grants	• revenue support grant • other grants from central government, eg for initiatives like e-government • generally, a substantial annual contribution to local authority revenue • of the order of 75% • small reduction in grant level has substantial effect in raising the need for council tax
Dividends and interest	• dividends received from investments and any joint venture companies • interest received from investments in bonds
European grants under the structural funds: • European Agriculture Guidance and Guarantee Fund (EAGGF) • European Regional Development Fund (ERDF) • European Social Fund • Financial Instrument for Fisheries Guidance	• 1 regions with lagging development • 2 regions (industrial decline — serious)) • 3 any area (unemployment — long-term) any area (excluded groups — integration) • 4 any area with industrial change (unemployment) • 5a agriculture (structural adaptation) • fisheries (structural adaptation) • 5b rural areas (vulnerable) • 6 regions (population — extremely low density)

Note: 1 to 6 denotes objectives relating to the types of region or area indicated — for which criteria are laid down for the receipt of funds

Organisation and structure for council tax

Although Index 1 (pp249–254) seems to suggest that more than 100 organisations, offices, officials and others are involved in the regimes for council tax and council tax benefit, the local heart of the system comprises three drivers, namely:

- the listing officer of the local office of the Valuation Office Agency
- the billing authority
- the local magistrate's court.

Listing officer

An area's listing officer (or assessor in Scotland) has five principal duties for council tax purposes, namely:

- to create and maintain the valuation list (see section 22 of the 1992 Act)
- to revalue property every 10 years
- to determine what is a dwelling
- to determine where cross-border property is situated for council taxation
- to determine the extent of a dwelling in a composite property
- to make and process proposals to change an entry in the valuation list (part of maintaining the latter).

The role of the listing officer is considered in greater detail in Chapters 14 to 16.

Valuation list

The assessments of dwellings, essentially for the subsequently authorised original valuation lists, were undertaken by the Valuation Office Agency under the Local Government Finance and Valuation Act 1991.

Subsequently, when council tax was first introduced, the listing officer for a billing authority's area became responsible for creating and maintaining the original valuation list by virtue of section 22 of the 1992 Act and is responsible for each revaluation (this was to be every 10 years under section 22B(3) of 1992 Act, as amended by section 77 of the Local Government Act 2003, but see p162).

Banding

Unlike business rates, which are based on a single national uniform non-domestic rate, the level of council tax is fixed for its area by the billing authority. Briefly, the amount paid for a dwelling (of similar value) differs in each billing authority of the three countries of Great Britain. It depends in part on the property's "band" (and exemptions, deductions and reliefs).

The listing officer allocated each property to one of a number of bands by reference to an estimate of its capital value. Unlike Northern Ireland which has discrete values for each property, the council tax bands are said to have a twofold advantage:

- they generally reduce the administrative load for the Valuation Office Agency and the billing authorities
- they reduce the likely number of appeals.

However, the IRRV Committee of Inquiry has questioned the use of banding. The legislation provides for the bands to be set differently in each country (for England and Wales see section 5 of the 1992 Act), ie one of eight old bands — bands A to H (on revaluation Wales gained band I). The assessed capital values for each property falls into its band of capital values — see Boxes 1.3 to 1.6 for the bands in England, Scotland and Wales (two boxes).

Taxpayers of properties in a band are charged council tax according to its band's allocated proportion of the band D amount — in accord with section 36 of the 1992 Act. (It may be noted that the Boxes 1.3 to 1.6 show the proportions of each band to band A in each box — to reflect the differences from band A to band H (or band I in Wales.)

Box 1.3 England — council tax bands and the proportions to band A

Band	Value as at antecedent date for England (1 April 1991) (£)			Proportion to band A
A	Up to		40,000	6
B	from	40,001 to	52,000	7
C	from	52,001 to	68,000	8
D	from	68,001 to	88,000	9
E	from	80,001 to	120,000	11
F	from	120,001 to	160,000	13
G	from	160,001 to	320,000	15
H	over	320,000		18

Box 1.4 Scotland — council tax bands and the proportions to band A

Band	Value as at antecedent date (1 April 1991) (£)			Proportion to band A
A	Up to		27,000	6
B	from	27,001 to	35,000	7
C	from	35,001 to	45,000	8
D	from	45,001 to	58,000	9
E	from	58,001 to	80,000	11
F	from	80,001 to	106,000	13
G	from	106,001 to	212,000	15
H	over	212,000		18

Box 1.5 Wales — old council tax bands and the proportions to band A

Band	Value as at the first antecedent date (1 April 1991) (£)				Proportion to band A
A	Up to			30,000	6
B	from	30,001	to	39,000	7
C	from	39,001	to	51,000	8
D	from	51,001	to	66,000	9
E	from	66,001	to	90,000	11
F	from	90,001	to	120,000	13
G	from	120,001	to	240,000	15
H	over	240,000			18

Box 1.6 Wales — new council tax bands and the proportions to band A

Band	Value as at the second antecedent date (1 April 2003) (£)				Proportion to band A
A	Up to			44,000	6
B	from	44,001	to	65,000	7
C	from	65,001	to	91,000	8
D	from	91,001	to	123,000	9
E	from	123,001	to	162,000	11
F	from	162,001	to	223,000	13
G	from	223,001	to	324,000	15
H	from	324,001	to	424,000	18
I	Over	424,000			21

Amount of council tax

The basic amount of council tax due on an individual dwelling (before considering discounts, exemptions and benefits) is calculated with reference to two key elements, namely:

- an assessment of the property's capital value band (see boxes above)
- the annual level set for the council tax which varies from area to area.

The assessment of capital values was undertaken by an independent government agency, the Valuation Office Agency (see above), and the annual level of council tax for an area is set independently by both billing and precepting authorities to meet part of their local funding needs for the year.

The assessment of capital value is to ensure that apart from being affected by funding requirements, the council tax due for a property is also related to the capital value of that property. To achieve this, capital values are expressed in terms of eight bracketed ranges, known as band A to band H (but Wales now has bands A to I). Whatever amount the billing authority wishes to raise from its district, the tax must also be set so that a property in band D gives rise to a basic council tax that is one and a half times that paid by a band A property and a band H gives rise to a council tax of three times that of a band A property and similarly accurate proportions for each of the other levels.

The capital value bands and the relative amounts of council tax imposed can be seen in Boxes 1.3 to 1.6 for England, Scotland and Wales respectively. (The last box shows the new band I for the revaluation in Wales.)

It may be noted that the relationships between the bands is usually expressed by reference to band D. Where band D is 9/9 the others are:

- A   6/9
- B   7/9
- C   8/9
- D   9/9
- E   11/9
- G   13/9
- H   15/9
- I   21/9

If the steps had not stopped at H (and now I in Wales), the amount of council tax paid by anyone with a house worth several millions of pounds would have been very substantial! As it is all extremely high value houses are in bands H and I (the latter for Wales).

Transitional arrangements

The new valuation lists result in many properties being allocated to new bands. In Wales, for instance, the 2003 revaluation resulted in:

- 53% going to higher band
- 1% to lower bands
- with 46% remaining in the old band.

Where the revaluation lifted a property's allocation to a band by two or more bands in Wales, regulations provide that a taxpayer who qualifies may receive "transitional relief" over the three years from 1 April 2005 (see Chapter 14).

Administration of local taxation

In the United Kingdom local taxation is similar in England, Scotland and Wales but there are differences of legislation and practice. (In Northern Ireland rating is still used for local taxation.)

Billing and precepting authorities

There are about 400 billing authorities and "major" and "minor" precepting authorities. The former, eg district councils, bill, collect and enforce council tax in Great Britain. Precepting authorities, eg county councils, call on the billing authorities to bill, collect and enforce the payment of council tax for them. Box 1.7 shows examples of the kinds of council tax billing and precepting authorities in Great Britain.

Box 1.7 Examples of billing and precepting authorities in Great Britain

Billing authorities

- England
  - metropolitan districts councils
  - district councils
  - unitary authorities
  - London borough councils
  - Corporation of the City of London

- Scotland
  - unitary authorities

- Wales
  - unitary authorities

Precepting authorities

- England
  - county councils
  - joint boards
  - town councils
  - parish councils (in general)
  - Greater London Authority
  - parish councils (area of the Corporation of London)
  - combined fire and rescue authorities
  - police authorities

- Scotland
  - community councils

- Wales
  - community councils

Setting of council tax

Section 33 of the 1992 Act and other enactments including the Local Authorities (Calculation of Council Tax Base) Regulations 1992 SI 1992 No 612 provide the basis for billing authorities to set the annual level of council tax. Once it is set, the level remains except for two possible reasons for change, namely:

- the government caps the level of the billing authority's council tax or that of one or more of the precepting authorities
- the level of council tax is challenged under section 66 of the 1992 Act and is quashed by the court on judicial review.

Capping of council tax

Where the level of council tax set by a billing authority or a precepting authority is considered too high, the minister may restrict or "cap" the level and require a lower level to be set. The power to do this is given by sections 30 and 31 and schedule 1 to the Local Government Act 1999. (The power to cap council tax is similar to the power to cap the old-style rating authority's rate poundage. This was given by the Rating Act 1984. However, with the national non-domestic rate (NNDR) there is no longer a need for non-domestic rating powers to cap business rates.)

Billing of tax

Every year, once the council tax is calculated by the billing authority, occupiers of dwellings are billed according to two factors:

- the local level at which the council tax has been calculated
- the dwelling's valuation list band.

However, some dwellings are exempt or attract relief (see Chapter 13). Similarly, some taxpayers do not have to pay tax or are entitled to a measure of relief. The tax is charged on a daily basis (see below).

Enforcement and counter-fraud work

When a taxpayer fails to pay the amount due, the billing authority will seek to enforce the payment under Part VI of the Council Tax (Administration and Enforcement) Regulations 1992 SI 1992 No 613. If necessary, action will be taken against the taxpayer (see Chapters 8 and 18). Similarly, where a taxpayer or a recipient of council tax benefit is believed to have committed an offence, eg under the Theft Act 1968 or the Social Security Administration Act 1992, the billing authority will investigate and, if appropriate, seek to penalise the alleged offender. Sometimes, the investigatory work will be undertaken with other bodies, eg the police or Royal Mail (see Chapter 10).

Best value and performance

Chapter 11 explores the application of best value and the assessment of performance in council tax matters. It may be noted that local authorities are "rewarded" in several ways by quality performance. Best value authorities may:

- receive best value grants (low tier authorities)
- achieve "Beacon Status" — as an exemplar authorities
- charge for discretionary services (see sections 93 and 94 of the Local Government Act 2003)
- trade through a company (see sections 95 and 96 of the last mentioned Act).

For the last mentioned, a new kind of company is now allowed, ie the "community interest company" (CIC), under the Companies (Audit, Investigations and Community Enterprise) Act 2004. The test for such a company is whether they exist for community benefit in the mind of a reasonable person. It is conceivable that such companies may be used in this context.

Taxpayers

Apart from occupiers of chargeable dwellings certain other persons may be liable, including:

- the personal representative of a deceased occupier
- the owner of a vacant property
- the occupier of the domestic part a composite property (or possibly the owner)
- certain other owners who are designated as liable instead of the occupiers.

Chapter 12 deals with taxpayers in detail.

Exemptions and reliefs

There are numerous exemptions and reliefs available which relate to certain groups of qualified individuals or to particular kinds of dwellings (see Chapter 13).

Daily charge

Section 2 of the 1992 Act provides that council tax is charged on a daily basis. Where, for example, the liable occupier has changed during the day, the circumstances which subsist at the end of a day are deemed to have subsisted throughout the day. In this instance, therefore, a new occupier who is liable and took possession at, say, 2 pm, would pay the council tax for the whole day.

Courts, tribunals and ombudsmen

The enforcement of the payment of council tax is a function of the billing authority. They may use the powers afforded to them to take taxpayers to court to obtain a liability order and then use the available remedies, eg bailiff action or attachment of earnings (see Chapter 8). The roles and activities of those in dispute are covered in Chapters 17 and 18, ie on the procedures of the courts, tribunals and ombudsmen.

Complaints

On matters of maladministration at both local and national levels of official administration the ombudsmen's services are available to deal with complaints (see Chapter 18). However, other procedures are available, eg a complaint to a local member of parliament.

Domestic rating in Northern Ireland

Northern Ireland has retained a form of domestic rating similar to that which had applied throughout the UK before the community charge was introduced. Box 1.8 shows features of the system.

Hitherto, rates in Northern Ireland were based upon the rental value of a dwelling at the relevant date. However, following a review in 2000, reform is progressing to change the system to one with the following characteristics:

Box 1.8 Types of domestic rates in Northern Ireland

Rates are charged in Northern Ireland	• not council tax
Two types	• district rates — fixed by the rating authority
	• regional rates — fixed by the Northern Ireland Executive or the Secretary of State for Northern Ireland
Valuations	• based on rental values (1976) but see below
Reform	• in progress following a review (see below)
	• revaluation to be based on capital values

- revaluation to capital values without banding (rather than the rental values of the 1960s as at present)
- transitional relief following the revaluation — over three years
- relief for disabled individuals — 25%
- relief for those on low incomes
- vacant property exemption
- payment deferment scheme for pensioners
- minimum payment requirement
- capping rates on higher value property
- changes to the appeals system
- a new valuation tribunal.

Revaluation

The last revaluation of the 700,000 domestic properties was in 1976 so the system has been operating for nearly 40 years. In 2006 a revaluation will be published in Northern Ireland by theValuation and Lands Agency with the view to commencing in April 2007.

A draft order to give effect to the above is set out for consultation (to end 30 December 2005) — it is the draft Rates (Capital Values, etc) (Northern Ireland) Order (2005).

Local Taxation — Principles from History

2

Aim

To explore the development of the principles of local taxation

Objectives

- **to review values and features underlying the principles of local taxation**
- **to explore the development of local taxation in the United Kingdom since early times**
- **to give a brief account of the search for additional or alternative taxes (see Chapter 19)**
- **to consider the progress of the Lyons Inquiry into local government funding**

Introduction

A debate on the future of local government funding in England commenced when Sir Michael Lyons began his enquiry in 2004 (see p224). Also, the other countries of the United Kingdom have reviewed or are reviewing their local tax regimes. It is timely, therefore, to review the principles of local taxation from the perspective of the history of domestic funding or rating.

The chapter begins with a consideration of issues and principles and then seeks to trace their development from early times. The possibility of change is always present so the following are considered in later chapters:

- improvements or changes to local taxation, ie additional or alternative taxes being introduced (see Chapter 19)
- improvements to the organisation of council tax (see Chapter 11).

Principles from local taxation

All taxes are said to be based upon established principles — an examination of general principles of taxation is introduced by an account of the developments of funding of what might be called "local administration". This account also attempts to compare historical features of local funding against the principles of taxation.

Box 2.1 lists the main values and features of local taxation from which the principles may be established. The box attempts to be comprehensive so that all forms of local funding or taxation may be measured against the principles. Note that some principles work in opposition to each other requiring a careful balance, for example, the possible trade off between fairness and simplicity

---

Box 2.1 Values and features of local taxation as a basis for principles

Fairness	• those of equal circumstances, status, income or wealth are treated equally
	• unequal treatments between those of different situations are appropriately justified
Efficiency of collection	• the proportion of tax collected is reasonable
	• the cost of enforcement is less than the arrears collected
Certainty or confidence	• taxpayers are sure that other taxpayers are meeting their obligations
	• proven offenders are duly treated
Neutrality	• the tax does not cause distortion in the local economy
Merit goods and services	• beneficial goods and services given adequate support
De-merit "goods" and	• the user pays
"services"	• user payment tends to be regressive
Local accountability	• elected or nominated persons are accountable for the administration and level of the tax
	• counter-fraud and counter-corruption measures are enforced
National accountability	• correct balance is sought
	• central government supports local administration and pays for "national" programmes and projects
Transparency	• there is freedom of information
	• there is participation by the public
Respect	• regard is had to human rights
	• data protection is observed, eg privacy
	• other measures of respect for the individual or family are observed
Representation	• those who pay taxes are adequately represented and consulted
Best value, performance	• the quality of services and goods is the best possible at the level of cost
and value for money	• the cost of collection is reasonable
	• value for money is achieved
	• continuous improvements are sought as a matter of course
External control	• the quality and other assurance mechanisms are sufficiently robust
	• they are open to external review and comparison
Audit and internal	• the internal operational systems are sufficiently robust to meet external and
inspection	internal review
Enforceability	• the non-payment of tax is resisted
	• costs borne by the non-taxpayer
Simplicity	• the regime is readily and easily understood
	• it is not unduly complex
Comprehensibility	• the regime seeks to cover all situations
Lawful practices	• practice complies with the law, ie it is intra vires
	• practice does not breach other bodies of law, eg human rights law

Early history of local taxation in Britain

The early development of taxation is misty, but could have developed from the principles of Roman taxation of 2000 years ago. Box 2.2 shows some of the key features of local taxation since Roman times. The box refers to "taxation" but this term does *not* imply cash payments in every period.

Roman taxation

The Roman era had two kinds of tax which were probably used extensively in Roman Britain, namely:

*   property tax
*   poll tax.

It should be noted that the property tax was charged on all asset-based wealth, including land. Land owners were required to register their land, so providing the authorities with a recorded tax base (akin to a valuation list, perhaps).

A feature of collection of taxes was "outsourcing". The authorities used contractors, the so-called "tax farmers", to collect taxes having paid for the right to retain the taxes they were to collect. However, the system was open to abuse and eventually abandoned.

Anglo-Saxon times

An important legacy of Anglo-Saxon times was the creation of an administration based on shires (except London) which were divided into hides. The geographical counties remained virtually untouched until 1974 when under the Local Government Act 1972 47 county councils were established. In Anglo-Saxon times units were the hide and the parochial parish.

It seems that there were only three public services due to the sovereign in Anglo-Saxon times, namely:

*   bridge building and maintenance
*   fortress building and maintenance
*   military services.

The king's treasury in Anglo-Saxon times, eg the reign of Alfred the Great, derived its revenues, both in kind and in cash, from a fairly wide base, including:

*   grants of land as bookland requiring services, including military service, bridge building service and fortress building service
*   grants of foodland requiring the payment of food-rent (food crops) or sums of money in lieu
*   sales, including produce from the king's estates, bookland and of rights in bookland
*   a miscellany, including tolls from bridges and fords tolls and fees for the dispensation of the king's justice
*   tributes, to pay Danegeld for instance
*   bequests.

Box 2.2 Profile of a few highlights of local taxation though the ages

Roman
- citizens were taxed on many forms of property and on the person
- assessments of land were undertaken by citizens registering their ownership
- taxation was often in the hands of "tax farmers"

Anglo-Saxon
- assessments of land were undertaken (essentially based on the hide, ie an administrative area)
- the owners of so-called bookland were charged three services to the monarch, ie fortress building and maintenance, bridge building and maintenance and military service
- the owners of foodland were charged annually to supply food to the monarch
- the land owners supported the church by paying tithes together with other payments
- the church or monasteries offered education and care services to the aged and sick

Norman
- William I introduced a feudal system
- land was granted to barons in return for military and other services
- barons made grants of land, and so on down
- the church maintained and strengthened its charitable role

Medieval
- auditor to exchequer first recorded in 1314
- poll tax imposed in 1377, 1379 and 1380
- result was the Peasants' Revolt of 1381 in Essex and Kent lead by Wat Tyler of Kent
- Statute of Cambridge 1388 made hundreds administrative areas responsible for poor — accommodation and care
- Statute of Mortmain 1389 required some of a benefice's assets to be distributed to the poor

Tudor
- with the dissolution of the monasteries in the 1530s, their almshouses and hospitals for the poor ceased as such
- Poor Law Act 1535 provided that the cost of "care" services fell voluntarily on local landowners — justices of the peace collected alms
- in the 1550s compulsory poor rates started in London and elsewhere
- poor relief started in various London hospitals

Elizabethan
- Poor Law Act 1562 provided for "voluntary" charity "rates" to be collected (failure to pay voluntarily resulted in an enforced charge, even imprisonment!)
- the Poor Law Acts 1593 and 1597 and 1601 resulted in a consolidated framework for poor relief at the parish level of Elizabethan society
- the Privy Council was responsible for the scheme

Post-Cromwell
- taxes were set annually
- money was raised three times a year
- assessments were undertaken, including land
- beating of the bounds (which has an ancient history) may have been used to identify or confirm land holdings for the purposes of assessment
- hearth tax introduced and repealed in the 17th century — many hearths bricked up to avoid the tax
- farming of hearth tax took place in some places
- window tax was introduced in 1696

Victorian era
- assessments were undertaken
- there were numerous bodies with power to demand rates
- district auditors appointed in 1840 to monitor Poor Law administration and rates
- 1851 window tax was abolished

Box 2.2 continued

20th century	• in 1925 the "general rate" was created by amalgamating various individual rates • assessments became a function of national administration in 1948 • rental rating of dwellings abandoned when community charge introduced • council tax brought in (with capital values of dwellings)
Current	• revaluations are in hand (England) or completed (Wales and Northern Ireland) • Scotland had a Bill to abandon council tax and introduce service tax (but it was not adopted) (see Chapter 19) • Lyons Inquiry considering the future form of local taxation

---

It is likely, however, that the church was growing in strength from these times and that the educational, medical and care services had their earlier roots then — the former was certainly supported by Alfred the Great. It seems possible that medical and care services were not of direct concern to the king. It may be noted that the church, particularly its monasteries, were concerned with medicine, eg herbs were grown for medicinal purposes, education and the alleviation of poverty. There was a second system of funding in that those with land were required to pay to the church tithes and a number of other "charges", eg for burials. It seems that the king's involvement in making grants of bookland was intended in some cases to enable the grantee to give support to the local church or monastery.

Norman "taxation"

The feudal system of administration created by William the Conqueror was initially service based with payment in kind with goods, labour or particular services. Later payments were commuted to rent or cash.

The exacting work required to create the *Domesday Book* was undertaken for several reasons. The time taken (two years) and procedures adopted were not unlike those for the modern revaluation. It involved the likes of:

- the identification of those who held land of the king and those holding land under them (a kind of hierarchy of occupiers in feudal times)
- the identification of property
- the description of habitations and other buildings and structures
- the creation of an inventory of animals and other resources owned by the occupiers
- finally, a partial or sample survey to check the original and to catch out any wrongdoers, an early scheme for "comprehensive performance assessment" perhaps.

A concurrent procedure was the assessment of property and assets for Danegeld (a tribute to stop invasion), which emphasises the model-like nature of Domesday for modern revaluations.

It was later that the notion of property rates was introduced, but this was at first a sporadic local impost in different parts of the country rather than a "national system" of local taxation. However, the tithe continued in Norman times and was in part at least, in effect, a form of local taxation to meet "social services" proffered by the monasteries, eg care of the sick and support for the impoverished.

Poor relief in medieval times

In medieval times much of the poor relief was undertaken by the monasteries but there were local rates levied from time to time. From the time of the demise of the monasteries, various statutes imposed the rates and other taxes, eg hearth tax. However, the Poor Relief Act 1601 (the Statute of Elizabeth) consolidated earlier legislation and introduced a rating system for parishes to raise funds for the relief of those in poverty. Initially, it covered more than real property but the general assessment of almost all personal assets began, through dispute, to take a limited form. For instance, the *Sir Anthony Earby's Case* 1633 resulted in the practice of excluding much of an individual's personal possessions, especially that which was held outside the parish — assessment was on more tangible, observable property. (The current debate on the future of council tax may require a fresh look at this result.)

Local taxation developed very slowly from this time and it was not until the 1800s that major changes begun the introduction of what became a national system by the latter half of the 20th century.

19th century developments

Under the Poor Law Amendment Act 1837 unions of parishes were created to administer the earlier legislation thus relieving the individual parishes of their duties. Abuses of the auditing of the unions' finances resulted, from about 1840, in auditors being officially appointed to monitor the way in which the Poor Law was applied by the unions. (The Select Committee on the Poor Law Amendment Act reported in 1838. The report included a recommendation that district auditors be appointed.)

20th century developments

A major reform of rating was introduced by the Rating and Valuation Act 1925. The Act provided for:

- local authorities to become rating authorities
- five-yearly revaluations
- assessment to be based, in essence, on rateable value
- valuations to be undertaken by the rating authorities.

The Local Government Act 1948 created a separation between the functions of assessment and raising local taxation. The Inland Revenue took over the assessment of property whereas local government authorities continued their responsibility for demanding, collecting and enforcing local taxation.

A major consolidation of the legislation was made under the General Rate Act 1967. A 1970 Act of the same title affected the rating of houses but it was not until the Local Government Act 1974 that the principle of rateable occupation was breached, ie when the rating surcharge was introduced to create the rate on vacant property.

Finally, at the turn of the century domestic rating was abolished and replaced with the community charge, the forerunner of council tax.

Review and change

Dissatisfaction with council tax is lively and in the last few years the administrations have instituted reviews at separate times (see Box 2.3). In Wales the revaluation which is now in force followed a review of council tax. In England and Scotland, council tax is under separate reviews and changes are possible to both. In fact the changes may be in abeyance in that:

- in Scotland persistent attempts have been made to introduce service tax (see Chapter 19) but these have been unsuccessful so far
- in England the government has announced a kind of moratorium on change — despite the continuing Lyons Inquiry.

Box 2.3 Recent reviews and changes to local taxation in the UK

England	Revaluation It was due as follows: • date of valuation 1 April 2005 • date of operation 1 April 2007 but has now been postponed Inquiry • review of local government funding by Sir Michael Lyons (see Chapter 2)
Northern Ireland	Rates Rates, not council tax, are charged in Northern Ireland in two forms: • district rates • regional rates Reform of the system is in progress following a review
Scotland	Review • an independent review was started on 16 June 2004 • Service Tax Bill promoted, but not adopted
Wales	Council Tax Reval 2005 • date of valuation 1 April 2003 • date of operation 1 April 2005 • transitional relief in place

---

Political futures for council tax

During the pre-election presentations for the election in May 2005, the three main political parties gave indications of their approach to the future of council tax. The principal proposals were as follows:

- Conservative Party — abolish the revaluation for England
- Labour Party — as part of the Budget

1.   pensioners to receive £200 rebate
2.   await the outcome of the inquiry by Sir Michael Lyons (see below)
- Liberal-Democrat Party — abolish council tax and replace it with a local income tax.

Additional or alternative local taxes

Several inquiries and other events suggest that local taxation never satisfies the taxpayers. A search seems to have been in hand for much of the time in the period from the late 1890s and there is the current inquiry by Sir Michael Lyons. A line of the major inquiries and other happenings are shown in reverse order in Box 2.4.

The above covered many possible additional or alternative types of local taxation. Chapter 19 reviews many facets of each of these and more.

Lyons Inquiry

Sir Michael Lyons is conducting an inquiry into local government funding, including council tax, and was due to report by the end of 2005. The Lyons Inquiry is likely to report on and possibly make recommendations on:

- the reform of council tax
- giving local authorities more flexibility to raise additional revenue
- the options for local income tax, reform of business rates and other possible ways to impose local taxation
- the implications for the financing of possible elected regional assemblies
- the impact of any changes to the rest of the UK.

However, in June 2005 the government indicated an intention not to change council tax nor introduce additional forms of local taxation! Also, the terms of reference for the Lyons Inquiry have been extended and it is now expected to report at the end of 2006.

Box 2.4 Inquiries into local taxation and some other events

Lyons Inquiry	• a current look into funding local government (see below)
Government	• White Paper Community Land Scheme • introduced from 1976 • objects of the community land scheme • part of the development land tax was to go to the local community • as land passed through public ownership (after the second appointed day), the land value accrued to the community • scheme was abandoned
Layfield Committee	• reported in 1976 • supported a local income tax • suggested a re-rating of agricultural property • recommended retention of domestic rating with valuation to capital values • dismissed site value rating, poll tax, payroll tax and several other
Government	• Green paper • published in 1971 • considered many possible additions to local taxes but did not go so far as to recommend any
Royal Institute of Public Administration	• tended to reject site value rating • reported 1956
Simes Committee	• reported in 1952 • considered site value rating only • majority of members rejected the tax
London County Council	• Bill on site value rating • published 1938 (but not adopted) • aimed to have the tax in London
Royal Commission on Local Taxation	• first reported in 1899 • showed systems did not have uniformity, equality and simplicity • recommended one valuation authority in a county • professional valuers should prepare valuation lists • properties should be valued once every five years • there should be one demand note

Part 2

Organisations and Structures

Government and Other Official Bodies

3

Aim

To describe the principal governmental bodies concerned with local taxation in the United Kingdom

Objectives

- to show how local government finance is raised
- to show the roles and activities of government participants in local government finance
- to identify roles and activities in the non-departmental governmental bodies concerned with local taxation
- to identify and outline the work of government departments concerned with local taxation

Introduction

Many government and non-government organisations, officials, offices and others have a direct involvement in council tax. This chapter is mainly concerned to identify and explain the work of the principal government organisations which have powers and duties for council tax. Also, it briefly considers the roles of official, independent bodies, ie non-government bodies such as the ombudsmen.

Local government finance

Funds raised for or by local government come from several sources, such as:

- national taxation and national insurance receipts
- borrowing by the issue of government gilt edge stock and municipal bonds
- revenue surpluses from property, business or commercial operations, eg rents, fees or commissions and sales
- capital receipts for the sale of land and other surplus assets.

(Box 1.2 in Chapter 1 gives a more comprehensive list of the principal sources.)
    Much of the money raised by central government is allocated to:

- the devolved governments
- government departments, eg HM Revenue & Customs, departmental agencies eg Valuation Office Agency and other non-departmental public agencies, eg Valuation Tribunal Service
- regional, county, district and other local government authorities
- European Commission (subject to the rebate to the UK) and other international bodies, eg the UN.

Issues on the allocation of centrally held funds to the regions, counties and other tiers of local government hinge on such matters as:

- the appropriateness of the formulae used to allocate funds to local populations
- accountability for funds which are not derived locally
- items which are considered by local communities to be national government responsibilities but are paid for by the local authority.

Although these matters are not directly relevant to the administration of council tax, they may have implications for it, particularly as locally raised funds are usually a tax on business and residential property occupation. In the case of business property, the amount raised in rates is £19 billion; council tax on dwellings raises £20 billion. Box 3.1 shows the place of local taxation in the context of the total amount collected as tax.

Box 3.1 Amounts of tax from the national and local sources of taxation			
	Tax	£ (billion)	£ (billion)
National taxes	• excise duties	40	
	• corporation tax	35	
	• income tax	128	
	• national insurance	78	
	• value added tax	73	
	• other taxes	62	416
Local taxes	• council tax	20	
	• non-domestic rates	19	39

Devolved government

The Scottish Parliament and the National Assembly for Wales have responsibilities for local taxation (see below). (In this respect the Scottish Parliament might one day abolish council tax and adopt a service tax but the Scottish Bill for this was not adopted earlier in 2005.)

Scottish Parliament

The Scotland Act 1998 gives the Scottish Parliament power to change local taxation in Scotland. If the Council Tax Abolition and Service Tax Introduction (Scotland) Bill of November 2004 had been passed by the Scottish Parliament, would have introduced a local income tax to replace council tax. The proposed date for the change was 1 April 2006.

National Assembly for Wales

Since it was formed under the Government of Wales Act 1998, the Welsh Assembly has exercised powers relating to local taxation including:

- a review of local taxation
- introducing the outcomes to the revaluation, eg transitional relief
- a reduction in the number of performance indicators used to assess the performance of local authorities
- the funding of the four valuation tribunals in Wales.

Government departments

Aspects of council tax, council tax benefits and other local taxes come within the remit of several government departments.

Main duties

A government department's main duties for the functions it deals with include:

- advising their secretary of state and other ministers
- liaising with the Scottish Parliament and the National Assembly for Wales, if appropriate, on matters involving common policy and practice
- taking and seeking advice on policy proposals, including proposals for EU policies
- drafting enactments, including those for EU directives, on government policy initiatives
- incorporating policies and practice on human rights
- the implementation of EU driven enactments
- the implementation of the law of freedom of information
- judicial reviews affecting the department's work.

The aggregate effect on local government is to seemingly increase the number and range of functions undertaken by the lower tiers of government. This increase may be funded by the departments concerned but any shortfall will be funded from local taxation or, if possible, other means.

HM Treasury

HM Treasury provides the link between the government's economic policies and the supervision of financial management of local government within the economy. It is the hub of the government's fiscal concerns, namely:

- to raise funds by national and local taxation, borrowing or other means
- liaise with other departments on expenditure plans, including those which affect local government
- to allocate funds to departments and others
- to ensure that the money is well spent
- to monitor and control the economy.

A government's first life is nearly five years from being first elected on the party manifesto, which sets out its aims and policies for the country. Implicit in the manifesto are a variety of programmes and projects which, if adopted by the government, will probably require legislation, and funding on a planned basis over the five years. The Chancellor of the Exchequer will need to manage the economy so as to meet planned expenditures; which will originally have been included in the planning phase on the basis of the Chancellor's expectations for the economy.

Each year the departments will have to "negotiate" with the Treasury for funds to meet its targets. Each year the Treasury will want "savings" and "best value" performance.

The Chancellor's annual budget cycle covers, broadly, three periods:

- from summer's end to the Budget Review Statement in late autumn (say, November)
- from late autumn to spring's Budget statement (say March or April)
- from spring to mid-summer's royal assent of the Finance Bill (say, July).

Each year the departments make their bids for funds and agree them with the Treasury. The Chancellor announces government expenditure, ie "total managed expenditure" (TME) in the Budget. (For 2005–2006 it is estimated at £548 billion.) The TME is split into:

- "departmental expenditure limits" (DELs) (say 60%)
- "annually managed expenditure" (AME) (say 40%).

Any DEL not used in the year may be carried over to the next year but not that in respect of AME. Thus, each department knows how much of its earlier bids will be available, together with any DEL carried over from the previous period.

Also, from the Budget Review Statement and the Budget, outsiders, eg local authorities, will learn of forthcoming allocations of government grants, new taxes, taxation levels and new or discarded exemptions and reliefs from taxation. Similarly, organisations will learn of new expenditure levels for their field and other matters affecting them, eg registered social landlords may expect £20 billion for new build housing in the three years to 2007–2008.

Her Majesty's Revenue & Customs (HMRC)

The HMRC is the government department with a remit among, other things, for overseeing the work of the Valuation Office Agency. Ignoring national taxation, local taxation covers council tax and non-domestic rating. As far as council tax is concerned the main concerns of the department are:

- advising ministers on the development of policy for local taxation in England and Wales
- overseeing the work of the Valuation Office Agency (see below).

Office of the Deputy Prime Minister

The Office of the Deputy Prime Minister (ODPM) has responsibility for local government and is, therefore, is concerned with several aspects of the council tax regime, including:

- policy developments
- translating policies into legislation, eg the local government enactments
- developing policy for best value
- overseeing the work of several inspectorates
- valuation tribunals.

Box 3.2 gives brief details of some of the local taxation policies and other matters which are the ODPM's responsibility. for capping the annual council tax proposals of billing authorities and precepting bodies.

Box 3.2 ODPM — selection of policies and activities on council tax

Capping of council tax	• the government has announced a level of 4% increase in council tax
Beacon Council Scheme	• several topics are listed for the Beacon Status Scheme
Local e-government	• IDeA is focusing on e-government for local government
Valuation tribunals (England)	• funded by the ODPM — 56 valuation tribunals in total

---

Department for Constitutional Affairs

The Department for Constitutional Affairs (DCA) is responsible for policy on matters relevant to council tax and council tax benefit, including:

- developing policy and practice on courts, valuation tribunals and other judicial bodies
- legal policy and the regime for enforcement
- the National Standards for Enforcement Agents.

The DCA is developing a major change in the way in which administrative justice is delivered and administered. It is anticipated that the various tribunals will be brought into a comprehensive service. At present it seems that the Valuation Tribunal Service will not be within the early part of the changes. The White Paper, *Transforming public services: complaints, redress and tribunals*, sets out the government's proposed reforms — they are somewhat wider than dealing with the tribunals.

Department for Work and Pensions

The department has several remits which impinge upon council tax and allied matters, including:

- developing social security and benefits policy
- handling claims for a wide range of benefits (other than council tax benefit which is dealt with by the billing authorities)
- overseeing the work of the Benefits Agency
- supervising the policy and work of the Benefits Fraud Inspectorate.

Department for the Environment, Food and Rural Affairs

The department retains an interest in council tax in so far as it concerns rural residential accommodation and agriculture.

Government agencies and other bodies

Government departments are the major components responsible for the delivery of government functions. In the main, the work is undertaken by the civil service hierarchy within the department. However, departments sponsor agencies, *ad hoc* committees, task forces, public corporations and other bodies. They may do this solely but sometimes with another department or even outside bodies. The agencies need to be considered in that they exist to support the work of a department and many have immediate relevance to council tax administration.

Many government operations are carried out by agencies rather than departments, the former being linked to the latter in a number of ways. Various terms are used to distinguish the agencies, such as "advisory", "executive" and "independent". The main agencies concerned with council tax and associated matters are considered below.

Valuation Office Agency

In England and Wales the listing officers of the Valuation Office Agency (VOA) were responsible for compiling the new valuation list for council tax for the areas of their billing authorities. They are also responsible for maintaining the lists, for the revaluations, and handling every proposal to change an entry in a valuation list.

For council tax, the VOA provides a legally binding and continually updated record of property units that are thereby designated according to law to be subject to the tax. This is, of course, subject to challenge in the courts by one or more parties, eg the owner or occupier, who may be aggrieved.

The database is used for other purposes and the VOA has other duties, such as:

- non-domestic rating
- valuation and estates work for government departments and many local authorities.

The VOA role and activities in council tax are mainly dealt with in Chapters 14 to 16 which concern the valuation lists and revaluation, the basis of assessment and some practical aspects of valuation for council tax.

Ordnance Survey

Traditionally, the Ordnance Survey makes the plans upon which the properties are shown (see Chapter 14 on matters of measurement for council tax). The agency is in partnership with other bodies to develop the National Spatial Address Infrastructure (NSAI) which will be available for a number of purposes, including:

- the administration of local taxes, eg collection of council tax
- an improved database of addresses for the purposes of council tax benefit.

Audit Commission

Under the Local Government Act 1993, the Audit Commission has the following main responsibilities:

- establishing and maintaining the principles and practice of best value
- overseeing the work of the Best Value Inspectorate (which incorporates the Housing Inspectorate).

In this respect it provides:

- general information on council tax practice
- best practice information
- audit reports.

Chapter 11 deals with the way in which the Audit Commission undertakes the many comprehensive performance assessments of units within the local government system and, briefly, its future.

Benefits Fraud Inspectorate

As an independent office of the Benefit Agency, the Benefit Fraud Inspectorate (BFI) has responsibility for the following main function:

- to fight any fraud and corruption in respect of council tax benefits (and housing benefits)
- to oversee the application of best value principles and practice in benefits administration to prevent fraud.

Chapter 9 covers the council tax benefits system which is administered by the billing authorities. The measures which may be used by the billing authorities and other bodies to combat fraud and corruption are dealt with in Chapter 10. Also, it considers, briefly, the future of the BFI.

Improvement and Development Agency

The Improvement and Development Agency (IDeA) focuses on local government with the intention of the encouraging the improvement and development of the best of public services. Its current key areas are:

- community well-being through engagement in local partnerships and other means
- e-government for delivering services
- improvement of the leadership in local service provision
- improvement of performance.

Local government

Local authorities — both billing and precepting

The council tax regime is administered by local authorities known as billing authorities (or levying authorities in Scotland). The roles and activities of the billing authorities and the precepting authorities are covered in outline in the introductory chapter and in several other chapters to which the reader may turn in due course.

   In terms of best practice and performance, they co-operate frequently with one another on such matters as:

- sharing information on best practice
- providing performance statistics for comparative purposes
- sharing information for counter-fraud purposes (see Chapter 10).

Joint boards

The joint boards were set up to administer some function of government which straddles the boundary of two or more local government administrations, eg some national parks have a joint board.

Police authorities

Certain police authorities are precepting authorities for council tax purposes.

Greater London Authority

By virtue of the Greater London Authority Act 1999, the Greater London Authority is the principal precepting authority for London. Its precept is billed by the London borough councils and the City of London Corporation.

Fire and rescue services

Section 3(5)(a) of the Fire and Rescue Services Act 2004 provides that a combined fire and rescue authority may issue a precept under section 40 of the Local Government Finance Act 1992.

Scottish regional authorities

On the original introduction of council tax in Scotland, assessors employed by the unitary authorities undertook the new valuation for council tax with guidance from the Valuation Office Agency.

However, council tax has not been abolished in Scotland and replaced by service tax in 2006. If this had happened there would not have been a need for a revaluation of the council tax in future.

Information Commissioner

Questions are frequently asked of the Information Commissioner about the secondary use of information obtained for council tax purposes. Such information must be used in a way which complies with the Data Protection Act 1998 and the Freedom of Information Act 2000 (see Chapter 5).

The Information Commissioner enforces the latter Act. A taxpayer who wishes to complain about access to certain information held, for instance by a local authority may make a complaint to the Information Commissioner (see Chapter 17).

Judicial bodies

The tribunals and courts that deal with council tax appeals are briefly described in Box 3.3. The three national ombudsmen who receive any complaints of mal-administration on council tax and council tax benefit are similarly covered below and in Chapter 17 which considers the resolution of disputes in detail and includes the main roles and activities of those involved in these bodies.

Box 3.3 Courts and tribunals concerned with council tax and council tax benefit

Benefit appeal service	• hears appeals on council tax benefit • with appeal on a point of law to the Social Security Commissioners (see Chapter 18)
Valuation tribunal	• hears contested proposals to change the valuation list
Magistrate's court	• hears cases on enforcement
County court	• hears certain matters from the magistrates' courts
High Court	• hears appeals from the valuation tribunals on points of law • with leave, allows application for judicial review • hears application for judicial reviews • may make certain forms of relief, eg an order for certiorari
Court of Appeal	• hears appeals from the High Court
House of Lords	• hears appeals from the Court of Appeal
European Court of Justice	• in principle, hears cases and appeals concerning European Union law • (none known on council tax and other local taxation matters)
European Court of Human Rights	• hears appeals on points of law concerning human rights

---

Judicial independence

As a matter of principle in the unwritten constitution of the UK there is a separation of the judicial function and executive function. The courts and tribunals concerned with council tax are dealt with in Chapters 17 and 18. In the latter chapter there is an analysis of the work of the valuation tribunals and the courts, together with that of support organisations, such as the Valuation Tribunal Service (VTS). The issue of judicial independence is touched upon in Chapter 18 in a brief resumé of the development and organisational structure of the VTS (see p208).

Ombudsmen for official work

The Parliamentary Ombudsman deals with complaints arising from the work of the civil service, ie on matters of alleged maladministration. Similarly, there are independent ombudsman services for local government and other public services in England, Scotland and Wales, namely

- the Local Government Ombudsman (England)
- Public Services Ombudsman (Wales)
- the Scottish Public Services Ombudsman.

The circumstances are usually cases involving complaints of alleged maladministration. For a review of the role of the ombudsmen, see p208. (The ombudsman for Wales replaces the Welsh Administrative Ombudsman and several other offices which deal with complaints in Wales by virtue of the Public Services Ombudsman (Wales) Act 2005.)

Professional Bodies and Other Organisations

Aim

To identify and describe the professional bodies involved in council tax

Objectives

- to describe the organisational structure and activities of a schematic professional body
- to outline the education, training and development of typical personnel in council tax
- to briefly describe the principal groupings of those involved in council tax
- to identify a selection of pressure groups concerned about council tax
- to show how individuals work with pressure groups or by other means to effect change

Introduction

Council tax, council tax benefit and other local taxation regimes have a cluster of administrators, professionals and others who devote all or most of their working lives to some area of administration, valuation or other aspect of the taxes. The area includes ancillary functions or services, with individuals who, for instance, sit in courts, enforce payment or teach. Many of these individuals belong to one or more professional bodies or other organisations, eg a Citizens Advice Bureau, with over-arching roles concerned with local taxes.

Professional body's functions

Although it has many very important functions, a professional body's principal role is developing its members' knowledge and capabilities by initial education and subsequent training; other functions support and extend this role for and, in effect, on behalf of the private or public client. The functions of a typical professional body are outlined in Box 4.1.

A professional body has several other functions, namely:

- as a builder of a professional ethos and morals for the protection of the public, and others

---

Box 4.1 A schematic professional body — principal functions and activities

Membership	• grades of membership • recruitment and induction of members • imbuing members with professional standards
Education	• liaise with academia at all levels • promote academic, professional and technical courses at appropriate levels
Training	• ensure members' early education is updated and extended, eg with continuing professional development and periodical publications • identify emerging training needs and meet them
Research	• identify research needs and priorities • identify academic and technical research requirements • encourage members and others to research
Information	• store and preserve professional information • give members, academics, students and others access to information • regular publication of information for members • publish research and technical data • provide forums and channels of communication between members
Standards of a technical nature	• formulate and publish technical standards of practice — mandatory and discretionary • liaise with relevant bodies on standards
Professional conduct	• formulate standards for professional conduct • publish and promote them to members and others, eg clients, government • cover such matters as: professional indemnity insurance (PII), clients' accounts, procedures for complaints against members and unbecoming professional conduct
Discipline of members	• set up disciplinary procedures • enforce the standards members are required to adopt
Promotion of the profession	• promote the profession to potential clients and other stakeholders, including the membership • maintain links with affiliated institutions • represent the profession to the media and the government

---

- as a learned body — being a wealth hoarder of professional history and knowledge which leads them to the generate research and contribute to education
- as creator of standards of professional and technical best practice, leading them to adopt, seek out or devise better methodology and apparatus
- as a disciplinary body to enforce professional probity and standards.

The council tax "industry" is directly and indirectly well endowed with administrative, professional or technical experts who cluster in one or more of several organisations.

Although a degree may be sufficient in terms of knowledge, most professionals gain credibility, and hence acceptability with clients and employers by being a qualified member of a professional body.

Roles and activities

The roles and activities dealt with here are rather different to those in other chapters since they concern the components of the structure of a professional body rather than a professional or other person. Most professional bodies are concerned with the matters shown in Box 4.1.

Administrators and property professionals

Traditionally, the word "professional" is either a noun or an adjective to indicate someone with a specific status with many constructs. In some ways, in modern times, it has become a cliché to denote any one who wishes to be so described. In local taxation the term is used to denote membership of a professional body which was formed in the past with the perspective, in part at least, to protect clients and the public at large from the mis-doings unprofessional of practitioners. Nowadays, membership requirements usually include:

- evidence of education or experiential learning
- qualification to undertake the work proclaimed
- willingness to undertake continuous development
- the carrying of professional indemnity insurance, if appropriate
- follow a code of good conduct and observe mandatory professional practice
- observe published professional guidance unless there are good professional reasons not to do so.

Individual bodies have different requirements but those given above are typical.

A particular body may have more than one grade of membership, eg "fellow", "associate member", "technical member", "probationer" and "student" or "pupil". Generally, much more detail of their work is given in other chapters.

The main groupings of professional administrators and property professionals involved in local taxation are described below, together with their principal professional bodies and a brief note of the kind of tasks they undertake.

In addition to the administrators and property professionals who work with governmental bodies, there are a number who have developed as specialists in judicial roles or academic roles, eg as researchers. Others are in a diverse range of companies and other organisations concerned with the likes of property and facilities management, counter-fraud control, performance, quality, and enforcement.

Roles and activities of professionals and others

Box 4.2 describes some of the many roles and activities of those public and private organisations, officers and others who participate in council tax and council tax benefit matters. Other chapters examine these and other roles and activities in somewhat greater detail.

Box 4.2 Roles and activities in council tax

Government departments	• (see Chapter 3)
• Treasury	
• Office of the Deputy Prime Minister	
• Department for Works and Pensions	
Government agencies	• (see Chapter 3)
• Audit Commission	
• Benefits Fraud Inspectorate	
• Valuation Office Agency (VOA)	
Listing officer (VOA)	• appointed for a billing authority's area
	• assesses or values properties for the valuation list
	• prepares the valuation list and publishes it
Assessor (Scotland)	• receives proposals to change bandings
	• values altered properties for a change of banding
	• attends appeals and presents the VOA case
Valuation officer	• listing officer or other officer of the VOA
	• undertakes valuations and other duties in that connection
Billing authority	• determines the amount overall amount to be collected including imposed precepts
	• determines individual liabilities
Levying authority (Scotland)	• bills, collects and enforces council tax
	• uses discretion to propose changes to a hereditament's inclusion in a band
Precepting authority	• determines the amount of its precept and informs the billing authority
	• receives amount from the billing authority
Rating surveyor	• advises owners and occupiers on the value of their property
	• if instructed, values and negotiates with the VO
	• if instructed, presents evidence at any appeal
Magistrate	• hears enforcement cases
	• issues liability orders
	• makes committal orders
	• hears appeals against distress
Civil enforcement agent (bailiff)	• following the making of a liability order, may execute a distress procedure
Valuation Tribunal	• hears appeals and decides the cases
Valuation Tribunal Service	• develops policy and guidance for valuation tribunals
	• delivers training programme for members
High Court	• hears appeals from the valuation tribunal on a point of law only
	• appeals go to the Court of Appeal and higher with leave

Career development

An individual seeking a professional career in local taxation may perceive it as developing through a series of notional stages, namely:

- part-time or full-time study to be eligible to qualify for professional status
- pre-professional development
- post-professional development as general practitioner
- possibly post-professional development to become a specialist (perhaps more than once).

While professionals are involved in practice in a functional capacity, say, as an accountant, lawyer or valuation surveyor, two other dimensions of their career may involve:

- knowledge and skills acquired and applied to various managerial functions
- knowledge and skills acquired and applied to various leadership roles.

Education, training and development

For those interested in a career in local taxation there are several routes. In addition to many employers, in general, three types of provider of education and training services may be identified, namely:

- further and higher education — the universities and colleges
- the professional bodies
- private sector training consultancies.

The Institute of Revenues, Rating and Valuation (IRRV) is very active in the training, education and development of revenues personnel. Among the resources and programmes offered are the following:

- education programmes leading to professional qualifications
- distance learning programmes
- a national on-line induction programme, called "Euclidian Benefits", for staff entering office concerned with housing benefit and council tax benefit
- a substantial number of specialist and general conferences and seminars
- links with local colleges for the delivery of professional education programmes.

As an illustration of the importance of training, some of the investigations by the Fraud Benefit Inspectorate suggest that new full-time staff should be recruited or part-time staff brought in to deal with large numbers of claims lack the necessary knowledge and skills.

Professional status

Each profession has its own routes to full practising status. Within a profession there may be more than one professional body, each with such requirements as follows:

- levels of entry

- pre-entry levels of educational attainment
- pre-practising attainments
- levels of membership
- continuing professional development standards
- certification to practice standards.

Both the IRRV and the Royal Institution of Chartered Surveyors (RICS) have programmes leading to specialist awards, eg the Rating Diploma of the latter which is essentially for rating practitioners. The IRRV's qualifications enable the holder to work in the local government sector, eg in the administration of council tax or council tax benefit administration (see Part 4).

Higher education

Entry with a university first degree in a cognate subject, eg estate management, is usual for most professions. Entrants with a non-cognate degree, eg a science degree perhaps, may enter some professions after or while taking a "conversion" course at post-graduate level. Full-time education at degree level is the most common mode of entry but part-time education suits many would-be professionals.

Part-time education may be by degree or some other acceptable mode of education, eg a professional body's own course and examination. Having obtained a first level degree and after some professional experience, many go on to master or diploma level study in specialist cognate subject. Others take a masters programme in business studies or carry out research for a master or a doctorate.

Probationer or trainee status

The steps to full membership of a professional body vary considerably between bodies. Typically, once an individual has completed a full-time cognate educational programme, probationer or trainee membership is usually available to the aspiring professional. Full membership will normally follow once a professional experiential programme has been completed to the satisfaction of the professional body. Such a programme might include:

- structured work experience
- a logbook which must be submitted to the body
- a written portfolio of work achieved
- an interview to discuss work done and knowledge of the profession.

Continuing professional development

Most professionals are required to undertake post-qualification training of at least, say, 30 hours a year. It is sometimes known as continuing professional development (CPD) and will cover the following:

- relevant, general studies or training to refresh, update, extend or maintain professional knowledge of everyday practice
- short-term specific studies or training to develop appropriately as a specialist in a new field — a common feature of developing for a "niche" market
- post-graduate study in the field or a field which is synergetic for a major personal development.

The above are not concerned with developing the individual in professional leadership or management but in "professional development". Of course some knowledge and skills may be transferable.

Management development programme

The principal role of the professional is, of course, that of adviser or expert practicing a profession directly to individual or to corporate or other clients in one of the property sectors. However, other less direct roles are within one of the following:

- a judicial or adjudicating function, eg an arbitrator, adjudicator and member of the valuation tribunal
- business managerial or administration function
- an education or training function, eg a teacher
- a research function, eg a researcher
- information or advisory function, eg a journalist
- evidential function, eg an independent expert witness.

Professional bodies

Institute of Revenues, Rating and Valuation (IRRV)

Unlike the RICS, the Institute of Revenues, Rating and Valuation (IRRV) is almost exclusively concerned with council tax, rating and associated local government matters. Membership comprises many of those in the public sector concerned with the administration of taxes raised for local purposes, particularly council tax. It is, therefore, the principal professional institute for council tax professionals. The IRRV provides training, professional support and technical and legal information and comment. It publishes professional books, journals and other publications, reports on case law, regulation updates and interpretations. Much of its material is available in digital form. It has reviewed local taxation with a Committee of Enquiry.

Chartered Institute of Public Finance Accountants (CIPFA)

The Chartered Institute of Public Finance Accountants is a professional institute providing technical information, analysis, audit guidance and comment on a range of council tax and council tax benefit matters. It also offers professional support, training, and information for public finance accountants and others.

Royal Institution of Chartered Surveyors (RICS)

The Royal Institution of Chartered Surveyors is the principal professional body concerned with the valuation of land and buildings. Professional activities involving council tax undertaken by those members include:

- making assessments of dwellings and other property for council tax

- making assessments of composite property for rating and council tax
- managing residential investment property, particularly houses in multiple occupation
- making determinations on appeals (as members of a valuation tribunals).

Rating Surveyors Association

Members of the Rating Surveyors Association are chartered surveyors who practice in the main on matters concerning non-domestic rating. Some will, however, be concerned with composite property and, perhaps less frequently, dwellings for council tax.

Association of Civil Enforcement Agencies

The Association of Civil Enforcement Agencies represents the companies which offer enforcement services to billing authorities and others. The Association provides the following:

- a code of practice which is mandatory upon its members
- training and education for enforcement agents
- publications
- procedures for handling complaints against its members
- a Professional Conduct Committee
- an independent panel to adjudicate
- consultation and advice to the government and other bodies on relevant topics, eg the National Standards for Enforcement Agents
- a complaints procedure.

The Local Authorities (Contracting Out of Tax Billing, Collection and Enforcement Functions) Order 1996 SI 1996 No 1980 enables a billing authority to contract out parts of its functions.

Advisory and pressure groups

Although mainly outside of government, many organisations advise or pressure national and lower tier governmental bodies to adopt, amend or drop policies, programmes or projects. On any concern involving council tax and associated matters, many managers, consultants and other professional advisers are likely to be working on one, two or more "fronts". A few of the possible pressure groups for an individual dealing with policy implications of council tax are shown in Box 4.3.

There are of course many owners, investors and others who raise issues and concerns in different ways on professional, trade or other matters, including:

- advising the senior management of possible implications for the business of the proposals
- advising clients of the opportunities and threats arising from the proposed changes
- seeking changes to the proposals by representing issues, problems, hardships and injustices to trade associations, professional bodies and to government officials
- drafting amendments to any Bill to allay inconsistencies, errors and omissions, eg by promoting exemptions and reliefs

---

Box 4.3 Some pressure groups with an interest in council tax and council tax benefit

Help the Aged	• has campaigned to change the way council tax impacts on older people's income • provides advice and information on reducing council tax
National Association of Citizens Advice Bureau (and local CABs)	• provides a comprehensive data source on, *inter alia*, council tax and council tax benefit • provides advice and information on reducing council tax and applying for council tax benefit
SPARS (Sparsity Partnership for Authorities Delivering Rural Services)	• has highlighted the disparity in levels of council tax between taxpayers in rural and urban areas
Council Tax Payers Association (Scotland)	• seeks replacement of council tax with another system

---

- if politically active, making representations directly to ministers or opposition speakers or by other means.

As an illustration, in 2000 the Social Security Advisory Committee on the Housing Benefit and Council Tax Benefit (General) Amendment Regulations (when in draft form) received representations from at least 83 local and national bodies.

Citizens' Advice Bureaux

Citizens' advice bureaux (CABs) offer a range of advisory services and in some instances advocacy roles, particularly concerning such matters as:

- information and proposals for resolving council tax arrears for individuals that seek their help
- providing demographic information such as deprivation indices
- assisting benefit seekers with applications for council tax benefit and the progressing their case.

Generally CABs are a source of feedback on perceptions, expectations, and difficulties facing individuals or communities who claim or receive benefits, or who are suffering hardship. Also, some billing authorities and their local CAB share information about council tax and other matters when dealing with claimants for council tax benefit — subject to the claimant's consent and other safeguards concerning electronic access for the staff of the CAB.

Information — Sources and Uses

5

Aim

To present sources of information on council tax and council tax benefit

Objectives

- **to identify roles and activities in organisations generating, gathering and promulgating information on council tax**
- **to identify the sources of information on council tax and benefit**
- **to explain the uses of information on local taxation**
- **to explain the principles of freedom of information in this field**
- **to outline the kinds of standards which impact on such information**
- **to provide a summary of information holders, gatherers and informants**

Introduction

This chapter is about the sources and the general uses of information which are available on council tax and associated matters. The gathering and kinds of information for the following purposes is dealt with elsewhere as indicated:

- information needed for valuation purposes (Part 6)
- information required for administrative purposes (Part 4).

Much of the information is in the public domain but not if it is protected under statutes, such as the Freedom of Information Act 2000 or the Data Protection Act 1998 (see pp50 and 52, for instance).

Roles and activities

Information on council tax and associated matters is generated, gathered and promulgated by a host of organisations and individuals in different guises. Box 5.1 attempts to conceptualise the roles into six generic groups, identify their specific roles and briefly indicate their activities.

Box 5.1 Information from conceptual and specific roles and activities involved in council tax and associated matters

Users	Clients	• to function as owners, occupiers and taxpayers concerned with dwellings and composite property
	Government (all levels)	• to develop and monitor policies on council tax • to develop initiatives
	Local government	• to administer council tax
	Professionals	• to advise clients on council tax • carry out work of valuing and managing property for council tax purposes • to advise on council tax benefit
	Journalists and mediumists	• to interest and inform sectional interests and the public on council tax matters
	Lobbyists and pressure groups	• to build a scenario or case • to inform for change in local taxation (or not)
	Students	• to learn about local taxes • to develop a specialism
	Teachers	• to inform • to develop a subject
Generators	Research students	• to achieve result knowledge • to develop a specialism • to obtain a qualification
	Researchers	• obtain new information and insights • to meet commissioned or supported output
	Opinion pollsters	• to establish held opinions • to inform
	Government	• to record of performance • to obtain statistics
	Analysts	• for analyses • to inform
	Map-makers	• to inform of place, direction and an area's features
Influencers	Government	• to commission • to meet needs • to analyse situations
	Research "commissioners"	• to obtain targeted research
	Professional bodies	• to meet members' information requirements about council tax • to develop a learned profession • to establish standards and guidance on best practice • to discipline members on professional conduct

Box 5.1 continued

Influencers	Research foundations	• to develop a learned society • to encourage researchers • to promote publications
Gatekeepers and controllers	Government (eg Office of Public Sector Information)	• to commission • to enable freedom of access • to keep secrets
	Clients	• to maintain confidentiality
	Professionals	• to maintain competitive position • to maintain commercial confidentiality • to promote business • to manage, design, construct and operate property
	Information Commissioner	• to protect personal data • to free information
Maintainers	Writers	• to advance recorded or known information • to establish linkages
	Glossarians	• provide insights • to inform meaning and remove confusion
Custodians and dispensers	Librarians	• to enable access and linkages
	Archivists	• to catalogue and store
	Information officers	• to enable access • to dispense • to inform

Uses of information for council tax

Statistical information is gathered for making policy on council tax, council tax benefit and for numerous reasons concerning government. However, the main uses of information for the administration of council tax are as follows:

- to identify the person who either occupies or owns a dwelling (see below)
- to determine whether a building is a chargeable dwelling or not
- to make assessments of dwellings for banding
- to determine a taxpayer's eligibility for exemption or relief
- to allow an application for council tax benefit
- to investigate and enforce, if appropriate, the payment of council tax or the repayment of council tax benefit in the event of suspected fraud.

Generally, personal and property information collected for council tax must only be used for the purposes and in accordance with Local Government Finance Act 1992 and other enactments. The Act sets out several supplementary sections on information (see sections 27 to 29, 38 and 52). Officers and councillors are required to ensure that the information is handled within the Act and complies with principles established in other statutes, eg the eight principles of the Data Protection Act 1998 (see below).

Enactments

Although the rating of domestic property under the General Rate Act 1967, as amended, was statute based there was relatively little secondary legislation when compared to council tax. The principal statutes for council tax are the Local Government Finance Act 1992 and the Local Government Act 2003 together with substantial volume of secondary legislation. (For these and other enactments which touch upon council tax see Appendix 2, Table of Statutes.)

Interpretation and definitions

A word or group of words used in an enactment are usually specifically defined within itself, another enactment or by reference to the Interpretation Act 1978. Otherwise, in a given dispute, the court will decide or give a meaning to the words in the circumstances of the case; on occasions, dictionary meanings give the meaning adopted or at least guidance. In some instances, the definitions of the words given in other statutes on seemingly unrelated legal matters are scanned for meanings or insights.

Circulars and other documents from government departments may be important statements of policy or guidance on the law but they are not authoritative interpretation — that is a matter for the courts. Thus, in *R v Maidstone Borough Council, ex parte Bunce* (1994) 24 HLR 375 circulars of the Department for Work and Pensions were not to be regarded as legal interpretation of the law.

Freedom of Information Act 2000

Generally, information is not freely available but the Freedom of Information Act 2000 (the 2000 Act) is an attempt to free information which does not need to be protected by secrecy. Thus, the 2000 Act provides members of the public right of access to information held by public bodies. Much information is readily available and some of the public bodies providing information about council tax are given in this chapter or elsewhere in the volume.

There are exceptions to the right of access, including:

- information which is barred by law, eg price sensitive information
- information on prospective policy or financial matters not yet determined by the public body
- information which would be expensive or unreasonable in personnel time or other costs
- personal information protected under the Data Protection Act 1998 concerning paper-stored and electronic-stored information respectively.

Sometimes a public body fails to supply or denies access to information. Following a direct complaint to the body concerned, a person who is not satisfied with the reason given, may make representations to the Information Commissioner who has been empowered from 1 January 2005 to determine whether the information should be made available.

Section 228(2) of the Local Government Act 1972 already enables a council taxpayer or others to see any order for the payment of work made by a local authority.

Codes of practice

Reference may be made to the defunct Code of Practice on Access to Government Information which laid down a number of exemptions from a general position that government held information was made available. In particular, Exemption 12 in Part II of the Code provided that the release of personal information was safeguarded in certain circumstances. Exceptions were available, for instance:

- where the court had ordered the release of otherwise protected information
- where the individual concerned gave permission for the information to be made available to a third party.

The Code has been replaced by the Freedom of Information Act 2000.

Some old cases before the Parliamentary and Health Ombudsman have involved complaints that property information held by the Valuation Office Agency (VOA) has been withheld from a payer of council tax seeking details of other properties. In some instances the VOA had written to the other parties seeking permission to release the information — some owners gave permission: others declined to do so.

The *Code of Practice on Consultation* is adopted by the government for its departments and agencies seeking views on a wide range of topics. Some six criteria are laid down, eg normally allowing at least 12 weeks for a consultation. The six criteria may be briefly described as follows:

- allow an adequate time for response
- give clear proposals, questions
- consult clearly
- provide feedback
- monitor the effectiveness of consultations
- follow best practice.

Property information

Although local authorities have power to gather information about property under section 16 of the Local Government (Miscellaneous Provisions) Act 1976, this is specifically prohibited regarding their requirements for information under Part I of the 1992 Act — paragraph 45 of schedule 13 to the 1992 Act contains the prohibition.

For the purposes of the original compiling of the valuation lists in Great Britain, a valuation of domestic properties was authorised by the Local Government Finance and Valuation Act 1991. Sections 4 and 6 of the Act empowered the collection of information about properties in England and Wales and in Scotland respectively.

National Spatial Address Infrastructure (NSAI)

The NSAI is being developed as a comprehensive database on individual properties from the following:

- the Local Land and Property Gazetteers (LLPGs)
- the National Land and Property Gazetteer (NLPG)
- the National Street Gazetteer
- the OS Address-Point
- the Postcode Address File (PAF).

Subject to agreement, the work is being undertaken by the Ordnance Survey with IDeA and others and completed within 18 months. It should then be available for council tax and other purposes.

Data protection

The Data Protection Act 1998 seeks to protect personal information held by employers, public bodies and other organisations. Stored information in hard copy and electronic formats must not be revealed to persons unauthorised by law to have access to it. Schedule 1 of the Act gives eight principles for the protection of personal data which must be observed. They are very briefly outlined here as:

1. fair and lawful processing
2. obtained for specified and lawful purpose(s) and not further processed in an incompatible manner
3. adequate, relevant and not excessive for the purpose(s)
4. accurate and kept up to date if necessary
5. kept no longer than necessary for the purpose(s)
6. processed in accords with the rights of the data subjects
7. protected by technical and organisational measures against mishandling, loss, destruction or damage
8. not transferred outside the European Economic Area unless the destination country has adequate level of protection and freedoms.

The Act provides for interpretation and for certain exemptions.

Law reports and interpretation

There are several general series of law reports which sometimes contain accounts of cases concerning council tax or council tax benefits. They include:

- Current Law (CL)
- The Times — reports in the newspaper.

Specialist series containing reports on local taxation including council tax and council tax benefits include:

- Rating Appeals (RA) — from 1961
- Rating and Valuation Reporter (RVR).

Many law books may be consulted and although they do not have the force or weight of a court or tribunal decision, they will usually give a good insight. They include the *Encyclopedia of Rating and Local Taxation* (see below regarding departmental circulars and other documents).

Government agencies and other offices

A key to the government information service is the Office of Public Sector Information on such matters as policy, standards and access. However, government agencies and other public offices issue circulars, reports, guides and other documents. While these are authoritative statements, like law books they are not to be taken as legally binding statements of the law — interpretation of the law is for the courts.

Box 5.2 shows the different kinds of information available from various government agencies and other offices (also see Box 5.1 above).

Box 5.2 Information provided by government agencies and other offices

Benefit Agency	• information on benefits made available to the public
Land Registry	• registers title of estates and interests • holds information about the title for public inspection in England and Wales • records the prices of transactions in the freeholds and leaseholds which have been registered. • publishes statistics on values (dated but it is available on-line)
Ordnance Survey	• surveys areas of new residential development and other construction • revises existing plans • produces new plans for various purposes and publishes them • with IDeA and others, developing the National Spatial Address Infrastructure (NSAI)
Valuation Office Agency	• created and maintains the original valuation lists • revaluation of domestic property (at 10 yearly intervals) • created the *Council Tax Manual* and the *Practice Notes* and maintains them • responds to enquiries form taxpayers and others in accord with the Citizen's Charter and the Turn Round of Post (TROP)

Ombudsmen

The offices of the ombudsmen each produce brief on-line accounts of their determinations in to alleged malpractice by billing authorities and others (see p208). On individual cases of malpractice the billing authority will receive a determination which is likely to point out deficiencies and those where there is a need for improvement. Many cases received by the ombudsman concern information about comparable properties which was needed for valuation purposes by a taxpayer appealing against a banding.

Local authorities

Each billing authority gathers information about occupation and ownership of dwellings and other chargeable property in their area. This information together with the valuation list bands and other information enables the billing authorities to charge those liable to pay council tax.

Many local authorities have websites which contain details of matters on council tax and council tax benefits as applied locally:

- information on exemptions and reliefs, eg council tax and students
- application forms, eg for council tax benefits.

Planning offices

The local planning offices hold any planning history about dwellings and other accommodation. Information on planning is normally supplied to the listing officer's office. However, in one situation the information was supplied at a late date and as a result the taxpayer claimed that a business property should be included in the valuation list for council tax: not the rating list for business property — he had bought not knowing of the situation.

Building control offices

The building control offices will normally hold plans and drawings of recently completed construction works, including new buildings and changes to existing buildings at the time of the application for approval of the works. It is important that the listing officer has information on changes to a building, eg an extension, so that the valuation list may be amended at the appropriate time, eg when the property is sold or at the next revaluation.

Information for other purposes

Paragraph 18A of schedule 2 to the Local Government Finance Act 1992 provides for regulations whereby authorities may use certain information in council tax administration which was obtained for other purposes — there are exceptions.

Standards, codes, notes etc

Several areas of administration are subject to standards, codes and notes. Many of these are for guidance but others are mandatory. For instance, the VOA deals with correspondence in accord with the standards of the Citizen's Charter and the Turn Round of Post (TROP). For the latter, 90% of enquiries should be dealt with within 15 working days and the remainder within a total of 20 working days.

Counter-fraud and enforcement standards

Much of the work in counter-fraud operations is bounded by standards laid down under enactments. Box 10.3 in Chapter 10 is an example which shows the extent of legislation in this area. Another example arises in the private sector. The National Standards for Enforcement Agents require certain levels of conduct in the work of enforcement agents. However, some billing authorities provide a form of "code" when an agent is engaged.

Measurement

The power of entry for valuing property, ie for the purposes of surveying and valuing whilst creating or maintaining the valuation list is restricted by section 26 (2) and (3) of the 1992 Act (see Chapter 16).

The *RICS Code of Measuring Practice* details the appropriate approaches to the measurement of dwelling according to the type of dwelling. Thus, flats are measured differently from houses for council tax purposes (see Chapter 16).

Council Tax Practice Notes

The Valuation Office Agency has published a series of numbered Council Tax Practice Notes, nine in total. Many deal with the detailed requirements of issues arising in valuation and allied property management matters. Box 5.3 gives the titles and an indication of the contents of each.

Box 5.3 Council Tax Practice Notes — numbers and titles

1   Definition of dwelling and basis of valuation for council tax
2   The valuation of composite hereditaments for council tax
3   The size, layout and character of the dwelling and the physical state of the locality — dates at which they must be considered
4   Disrepair, building works, temporary disabilities
5   Dis-aggregation of dwellings
6   Premises in multiple occupation (aggregation of dwellings)
7   Application of council tax to caravan pitches and moorings
8   Domestic/non-domestic borderline
9   Cross boundary dwellings

Journals and other publications

A number of publishers and professional bodies offer journals, books and CD roms which cover council tax and council tax benefits in a general or specialist manner. Examples of journals include:
* the *Estates Gazette*
* *Insight* — published by the IRRV.

National statistical returns

The Chartered Institute of Public Finance collect data on council tax from a number of sources. The data is collated and the information is published either as comparative statistics or as other kinds of report.

Internet and the world wide web

The internet and www is increasingly relied upon as a link for information, services and, indeed, a host of operations concerning council tax, including

- telephone number listings
- property listings
- insolvency practitioners
- valuation office
- appointed bailiff contractors
- legislation database providers
- best practice database providers
- case law database providers
- professional institutes
- ODPM
- Her Majesty's Stationary Office
- banks
- direct debiting systems and other financial operations.

Billing authority websites

Many local authorities have websites which contain details of such matters as:

- council tax, eg transitional relief in Wales
- council tax benefits
- exemptions, reliefs and reductions, eg council tax and students
- counter fraud policy and practice.

Powers to seek information for council tax purposes

The enactments provide for many bodies to seek information so as to enable them to fulfil their duties under the legislation for council tax and council tax benefit, including:

- the listing officer (see Chapter 16)
- the billing authority.

Valuation

Each billing authority has a valuation list for its area. It is created and maintained by the listing officer of the Valuation Office Agency. The listing officer has powers under the 1992 Act, namely:

- to enter a property to obtain information for valuation purposes (section 26(1))
- to obtain information about a property (section 27).

In the buying and selling of a dwelling, a buyer or his or her solicitor is required under the Finance Act 1931, section 28, to supply HM Revenue & Customs with certain details of the transaction, the "particulars delivered", particularly for stamp duty land tax purposes. The information is recorded on the database and may be used for banding purposes.

Administration

In the administration of council tax, billing authorities have powers to seek information for a number of purposes, including:

- in billing, occupier's or owner's details
- in enforcement, income and employment details.

A billing authority is empowered to collect information which will enable it to identify who is liable for the council tax for a dwelling. This is done under Part II of the Council Tax (Administration and Enforcement) Regulations 1992 SI 1992 No 613 (see Chapters 8 and 12).

Under regulation 3 the billing authority may do this by serving a notice on a person, eg a resident or property manager, thought to know who may be liable. Within 21 days the recipient must supply any information within his or her knowledge. Information may also be sought from official bodies under regulation 4 of SI 1992 No 613. However, a police authority (as such) or an employer must not be approached for information in this way.

By way of a summary, Box 5.4 gives the official bodies who may supply information to a billing authority for the purposes of council tax and council tax benefit.

The registrar of births and deaths is required to notify the billing authority of the death of an adult in its area within seven days of the death. This may be done by, say, a simple printout of the list of deaths for the registrar's area being dispatched to the relevant billing authorities.

Box 5.4 Official bodies who may be asked to supply information

Precepting authorities	• may be asked to supply information
Other billing authorities	• may supply information or exchange information
Electoral roll officer	• supplies information about individuals on the electoral roll to the billing authority • does not supply those on the postal list
Registrar of births, marriages and deaths	• supplies information of deaths to the billing authority

---

Demand for income and employment details

A billing authority may demand information on income and employment after a liability order has been granted to the billing authority in respect of a debtor. Details of this remedy for enforcement are given in Chapter 8 (see p92).

Promotion of revenue services

One of the uses of information is in the promotion of revenue services by a billing authority. "Promotion" is more likely to be considered in the context of the private sector rather than in the context of public revenue services, such as council tax and council tax benefits. In revenue services, promotion may be considered as the process of interaction between the provider, ie the billing authority, of the service and one or more of the following:

- the payers of council tax
- recruits to the revenue service who need induction
- claimants or prospective claimants of council tax benefits
- after elections, recently elected councillors who are new to the full scope of revenue services
- taxpayers, claimants and others who are thought to have committed fraud
- existing staff of the revenue service.

The interaction may comprise:

- encouraging defaulting taxpayers to pay
- informing the applicant or the taxpayer of the alternative methods of payment
- encouraging the take up of the direct debit payment method
- encouraging staff to undertake training and development
- informing prospective claimants of their entitlements
- eliciting information about status regarding exemptions, deductions or reliefs
- inducting new councillors and newly appointed staff
- as necessary, progressing an enforcement or recovery case in the interim stage of a hearing.

Promotion — purpose or objectives

When a revenue service promotional strategy is developed it should have a mission and stated objectives. Any subsequent promotional activities should be identified against one or more of the objectives. Promotion is intended to inform and motivate those who are or should be affected by the service, for example:

- of the many attributes of the service — a basic information to a largely undifferentiated audience
- that the bundle of attributes suits the applicant's requirements — an emotional appeal perhaps
- of the key features of the process (as conducted by the service) — creating an understanding of the process for say a person facing a hearing before a magistrate, eg the enquiries stage (see Chapter 18)
- the positive nature of any endorsements of the service, eg beacon status — again, an emotional appeal for "third party" support

- meeting of targets and good performance, eg informing stakeholders by the best value performance plan (BVPP) (see Chapter 11).

Promotional mix

Possible messages to the target audiences for revenue services may be delivered by a typical promotional mix as shown in Box 5.5. The box covers typical private sector approaches which might be adopted by revenue departments and in-house initiatives developed by various offices. Each audience will or should receive message(s) identified as appropriate for it.

Box 5.5 Possible profile of a promotional mix for revenue services

Element	Medium	Target audience
Advertising	• paid advertisements placed in newspapers, magazines, and journals or on buses and at stations	• general public • segment of public • professional group(s)
Publicity	• press releases used without advertising support • free editorial and features about the revenue services placed in newspapers, etc and other material which is distributed in the market	• public or segment of the population • professional audience (or segment) (message will depend on the audience of above)
Managed events	• information distributed at exhibitions, conferences and other events by speakers, brochures, leaflets and poster	• attendees • specialists • recruits • taxpayers (messages may be educational, political, motivational etc or a mix)
Personal selling	• details given by word of mouth to key market "gatekeepers" of information, eg press representatives	• all involved, but message is designed to audience
Special programmes	• benefit surgeries • officers as home visitors with PDAs for processing applications for council tax benefits • benefits road shows and benefits buses • best value performance plan (BVPP)	• claimants and "hidden claimants" • above and benefits staff • above and advisors • all stakeholders

---

Target audiences

It is not sufficient to prepare the promotional mix and hope that it has the desired effect. A target audience will need to have been identified in terms of each message or messages that it is intended to deliver to them. One categorisation of a "market" for a service is as follows:

• buyers — those who pay for the service
• users — the persons who will use the service, either as taxpayers or claimants but also as management and staff of the service or councillors
• influencers — those who, say, support or advise the users, eg councillors and CAB staff
• deciders — on the person or persons who make the decision to buy the service, eg councillors.

It follows that the promotional mix will not be the same in terms of the information content, slant and emotion for each target audience. For instance, the targets, message content and timing will be very different for the following:

• seeking to increase the take up of council tax benefit
• seeking to increase the take up of the direct debit payment method.

In this context revenue officers need to be aware of the right to privacy and other rights that taxpayers and others have in this context.

Information holders, gatherers and informants

The administration of council tax and the compilation and maintenance of valuation lists requires substantial databases. Box 5.6 summarises some of the information given in other chapters. It shows in a general list information holders and gatherers who are informants or may be called upon to become informants.

Information is protected in that:

• its use must be sanctioned under the 1992 Act and other "operational" enactments
• gathering and promulgation must comply with the requirements of the
   1. Data Protection Act 1998
   2. Human Rights Act 1998
   3. Freedom of Information Act 2000
   4. the investigatory statutes given in Box 10.2.

Holders, gatherers and informants, as well as official users must be circumspect in their activities; for example, the Benefits Fraud Inspectorate has pointed out the non-compliance of some of the latter.

Box 5.6 Information holders, gathers and informants

Source	Information provided
Registrar of births marriages and deaths	• supplies information to help maintain the database of liable parties and liabilities
Council tax payer	• supplies information to help maintain the database of liable parties and liabilities
Estate agent	• supplies information to help maintain the database of liable parties and liabilities • provides information to assist in revaluation
Letting agent	• supplies information to help maintain the database of liable parties and liabilities
Solicitor	• supplies information to help maintain the database of liable parties and liabilities • supplies transaction details, eg stamp duty land tax return
Electoral registrar	• supplies information to help maintain the database of liable parties and liabilities (although less so now that there is an option for voters to be on the non-public list, which is not available for the purpose of council tax registration
Tracing agent	• provides forwarding addresses for absconded liable parties
Inspector	• employed inspectors identify completion date of properties and liable parties
Bailiff	• provides information about defaulters they have visited, including employment details, distrainable goods, and financial status
Benefit matching systems	• national databases that provide limited data to assist in the detection of fraud by identifying mismatched records
Planning department/ building control departments	• provide information about new builds and property improvements
Street naming and numbering unit	• provides information to help maintain the council tax properties data base
Benefits agencies	• passes on council tax and housing benefit claims made at their outposts • provides information to verify council tax benefit claimants status • provides details of commencements and cessations of direct deductions from income support or job seekers to pay council tax arrears
Employer	• informs the local authority if a person subject to a council tax attachment to earnings order leaves their employment
Citizens Advice Bureau	• provides information and proposals for resolving council tax arrears for individuals that seek their help • provides demographic information such as deprivation indices • provides feedback on perceptions, expectations, and difficulties facing individuals or communities who claim or receive benefits, or who are suffering hardship
Insolvency practitioners	• provide information on those who are or are about to enter one of the variety of types of insolvency or bankruptcy
Banks	• provide confirmation of direct debit collection and details of failed direct debits
Valuation Office Agency	• provides a legally binding and continually updated record of property units that it has designated according to law to be subject to council tax

Box 5.6 continued

Source	Information provided
Other local authorities	• share information on best practice • provide performance statistics for comparison
Internet world wide web	• increasingly relied upon as a link for information, services and indeed operations including telephone number listings, property listings, insolvency practitioners, valuation office, appointed bailiff contractors, legislation database providers, best practice database providers, case law database providers, professional institutes, ODPM, Her Majesty's Stationary Office, banks, direct debiting systems
Audit Commission	• provides general information • provides best practice information • provides audit reports
ODPM	• provides operational news letters • copies of press releases • consultation papers and their results • regulation proposals • regulations becoming effective • special reports effecting local taxation usually commissioned by central government • statistical data • best value data • comprehensive performance assessment data • a variety of other information both about and for local government
Institute of Revenues, Rating and Valuation	• the professional institute for council tax professionals • provides training and professional support • provides technical and legal information, comment, and professional publications, case law and regulation updates and interpretations
Chartered Institute of Public Finance Accountants	• professional institute providing technical information, analysis, comment, audit guidance • provides professional support, training • provides other information for public finance accountants and others
Law reporters and publishers	• case reports and databases of cases reported

Part 3

Property and Management

Property Covered by Council Tax

6

Aim
To identify dwellings and other property for which a charge to council tax arise

Objectives
* to describe the types of property which are statutorily identified for council tax
* to explain the nuances of such property in terms of the occupier's or owner's liability for council tax
* to identify exempt dwellings
* to explain the issues of working at or from home

Introduction
This chapter describes the types of property in respect of which liability arises — mainly on the occupier. Not only does it cover occupied property but also unoccupied and vacant dwellings. Also, certain types of dwelling and other accommodation result in the owner being liable, rather than the occupier or resident (see *Prescribed dwellings* below).

However, certain properties may have no tax, reduced or discounted tax — the following are dealt with elsewhere in this volume:

* the concept "sole or main residence" of the taxpayer — this issue is dealt with in Chapter 12
* exemptions, deductions, discounts and reliefs afforded to certain individuals is covered in Chapter 13,

Although most dwellings are houses or flats, a miscellany of other less usual properties are caught, eg caravans and houseboats. However, some dwellings are not covered by council tax, eg some holiday cottages — they are treated as non-domestic property subject to business rates.

The topic is quite complex and has a multitude of nuances which must be borne in mind when considering a type of property. Thus, informally several main groups of dwelling may be identified for council tax, namely:

- ordinary houses and flats
- granny flats and similar annexes
- certain types of short stay accommodation (see p182)
- caravans and houseboats
- certain prescribed dwellings (see Box 6.4)
- the domestic component of composite property, ie property which has a mix of non-domestic and domestic accommodation, such as the domestic accommodation of a public house
- dwellings that are exempt
- second homes.

By way of summary in this context, "domestic property" means a unit of property that is not shown on any non-domestic rating list. In effect houses, flats and other property which are not used for residential purposes but are wholly used for business or other non-domestic purposes are not within the ambit of council tax.

Chargeable dwellings

Formally, the legislation gives a multi-faceted, round-about-way of defining "chargeable dwelling" including that within composite property; using the following enactments:

- section 3 of the 1992 Act which takes the concept of the "unit" of section 115 of the General Rate Act 1967 (resurrected for the purpose)
- sections 66 and 66A of the Local Government Finance Act 1988
- "not non-domestic" from section 2 of the 1992 Act.

Appendix 6 briefly examines these statutory provisions. Although section 3 of the 1992 Act gives the meaning of "dwelling", the Secretary of State has power, by order, to amend any definition of "dwelling" (see section 3(6)). Thus domestically occupied houses, flats and the like are covered by council tax unless an exemption applies. Also, the part of any business property which has some residential accommodation, eg a caretaker's flat, will be caught for council tax as the residential element of a "composite property" (see section 3(3)). Section 7 of the 1992 Act provides for liability in respect of caravans and boats — with special meanings for "caravan" and "owner".

Case law

Case law interprets the concept of "dwelling" for council tax purposes and a selection of cases is given in Box 6.1. It may be noted that *Batty* v *Burford* [1995] RA 299 dealt with the question of whether a granny annexe was a separate dwelling for the purpose of listing in the valuation list. The separate issue as to whether a particular annexe type dwelling is subsequently exempt from council tax may be influenced by it being prevented by law from selling or letting separately (see Chapter 13 *Exemption: Class T*).

Gardens

Where a garden, outhouse or other appurtenances are with a house, section 3(4) of the Local Government Finance Act 1992 provides that it is included to the extent that it may enhance the value

Box 6.1 Cases on the concept of "dwelling"

Houseboat and the like | *Nicholls* v *Wimbledon Valuation Office* (LO) [1995] RVR 171
Held: a floating house which has no means of propulsion may be a chargeable dwelling

Stubbs v *Hartnell (Listing Officer)* [2002] RVR 90
Held: a houseboat at a mooring together with a plot of land was a hereditament, ie a separate chargeable dwelling, for council tax purposes

Caravans | *Field Place Caravan Park Ltd* v *Harding (VO)* (1966) 196 EG 469
Held: caravans on a site were held by the owners and not the site owner, they were enjoyed with the land and were sufficiently permanent

Boardman v *Portman* [2001] LGLR 7
Held: two weeks occupation of a caravan was insufficiently permanent, ie it was transient — the occupier was waiting to move into a house.

Divided or disaggregated property | *McColl* v *Subbachi (LO)* [2002] RVR 342
Held: where accommodation, a flat in a house, was accessible though the house and was separated by a lockable door, the two comprised separate dwellings

Granny flat | *Batty* v *Burfoot* [1995] RA 299
Held: granny flat or annexe — incorrect to consider the degree of communal living or if the annexe could be sold separately — the issue is construction as a self-contained living unit

Flats or flatlets | *Beasley (LO)* v *National Council for YMCAs* [2000] RA 429
Held: for a property of 10 flatlets it was the construction rather than the use which determined their tax status

Rodd v *Ritchings* (1995) The Times, 21 June p22
Held: a property divided into two or more separate dwellings may result in more than one chargeable dwelling

Beach huts | *Lewis* v *Christchurch Borough Council* [1996] RA 229
Held: beach huts were separate dwellings for council tax purposes — they were:
• self-contained
• not non-domestic properties

Showhouse | *Fawcett Properties Ltd* v *Buckinghamshire County Council* (1961) 176 EG 1115
Held: a show house is non-domestic

Walker (VO) v *Ideal Homes Central Ltd* [1995] RA 347
Held: a showhouse is a non-domestic hereditament resulting in a liability to rates, not council tax

of the house, even one on the other side of a road (the right to use the garden is not regarded as being taxed as such). However, a separate garden some distance away would not be included (and would be outside of the council tax regime).

Short stay or self-catering accommodation

The provisions of the non-domestic rating Act, ie the Local Government Finance Act 1988, cover short stay accommodation and self-catering accommodation in subsection 66(2) and 66(2B) respectively. The subsections set out criteria for such property to be non-domestic. In brief the criteria in section 66(2) for short stay accommodation are that it is:

- wholly or mainly used in the course of business for short stay accommodation
- provided for short periods to individuals whose sole or main residence is elsewhere
- it is not self-contained self-catering accommodation
- it will not be offered to more than six persons simultaneously.

The other subsection covers the following:

- it is a building or self-contained part of a building
- it is available for commercial letting
- it is to be let for short periods for self-catering purposes
- the periods are intended as short periods totalling 140 days a year.

Finally, short stay accommodation, eg a hotel, is normally excluded from "domestic" but some issues arise as to whether this is so in every instance. The accommodation covered includes:

- hotels
- guest houses
- bed and breakfast accommodation
- holiday cottages.

Such accommodation may be seasonal or the taxpayer may occupy almost all the accommodation for living purposes. Thus, some such accommodation may fall into the council tax regime because the short stay use is merely subsidiary to the residential occupation by the taxpayer or is the domestic element of a composite property (see below). Box 6.2 summarises aspects of short stay accommodation.

Second dwellings

The White Paper of 1991, *A new tax for local government*, suggested that a second dwelling (a somewhat colloquial term) should result in a bill for local tax. A number of issues concerning second homes exist, namely:

- like other homes in their area, result in costs to the local authorities
- demand for services is unevenly distributed over the course of the year, eg it is higher in, say, the tourist season

Property Covered by Council Tax

Box 6.2 Short stay accommodation and council tax

Hotels	• normally charged to non-domestic rates (see section 66(2) of the LGFA 1988) • guests have main residence elsewhere • council tax is billed where accommodation is occupied by the owner, guest or a member of staff, ie as their sole or main residence
Hostels and like	• some hostels for the homeless and others are exempt from council tax (see below *Exempt Dwellings*)
Holiday cottages, blocks of flats and other similar property to let	• must fit the criteria of section 66(2B) of the 1988 Act — non-domestic rates are charged
Bed and breakfast accommodation	• council tax billed where the short stay element of accommodation is no more than subsidiary to the taxpayer's residential accommodation
Timeshare	• if timeshare accommodation is within the Timeshare Act 1992, it is not domestic (see section 66(2E) of the 1988 Act)

Box 6.3 Second homes and the council tax percentage discount

England	• discount of from 0 % to 50 % available to second home owners • at the discretion of the billing authority
Wales	• discount of from 10% to 50 % available to second home owners • at the discretion of the billing authority
Scotland	• a 50% discount available to second home owners

- unoccupied second homes need certain services, eg fire and emergency
- in attractive tourist areas, they reduce the availability of dwellings to local home buyers — mainly because the prices tend to rise.
- in some areas the authorities have introduced a habitation condition — new dwellings are only available to those who meet residential qualifications.

As a result section 11A(3) of the Local Government Finance Act 1992 (as inserted by section 75 Local Government Act 2003) provides for the billing of second homes but the circumstances need to be considered on a situation by situation basis. For instance, dwellings offered as holiday cottages or caravans and the like are normally subject to business rating — certain conditions need to be observed.

However, where the second home is truly a home — not a business — council tax is charged as shown in Box 6.3. It may be noted that a second home is defined as a furnished dwelling that is nobody's sole or main residence for up to six months. This applies to furnished lets during unoccupied periods. The reduction in discount will not apply if the liable person is paying council tax on another property at which he must live for purposes of his employment.

An unfurnished dwelling will be exempt for a period of up to a maximum of six months.

Once a dwelling has not been anyone's sole or main residence for more than six months, it is known as a "long term empty property", for which the billing authority may, at its discretionary policy, reduce the discount from 50% to as low as nil.

Other dwellings

Certain other types of residential accommodation are caught under the 1992 Act, including:

- a farmhouse and workers' cottages (although agricultural land and buildings, ie those that are not dwellings, are exempt from business rates)
- dwellings within a non-domestic premises, eg living accommodation in a public house
- certain dwellings within a business valued for non-domestic rates by the formula approach.

Aspects of the valuation of some of these dwellings are dealt with in Chapter 16.

Prescribed classes of dwellings

Under section 6 of the 1992 Act certain properties containing residential accommodation are by regulations a prescribed class of dwelling and the owner is liable for council tax. The types of property are shown in Box 6.4.

Class C: houses in multiple occupation

The wording covering Class C has been interpreted in several cases. For instance, in *Hayes v Humberside Valuation Tribunal & Kingston upon Hull City Council* [1997] RA 236 the Court of Appeal upheld the valuation tribunal decision that the dwelling was an HMO because the fitting of separate security looks on some of the rooms constituted an adaptation to accommodate persons who do not occupy a single household. In *UHU Property Trust v Lincoln City Council* [2000] RA 419 it was found the realities of tenancy/licence arrangement (not just the written agreement) must be considered in determining whether a house is a HMO (see Box 6.4).

Class F: accommodation for asylum seekers

Section 95 of the Immigration and Asylum Act 1999 provides that the owner of accommodation provided for asylum seekers is liable for council tax, not the resident asylum seeker.

Composite property

Composite property is a hybrid property comprising part non-domestic and part domestic. The non-domestic part is covered by non-domestic rating whereas council tax is chargeable in respect of the domestic part. The main issue is whether the accommodation is used only for residential purposes.

Box 6.4 Occupied homes giving rise to non-resident owner's liability

Class A	Residential care home	• nursing home or mental nursing home (Registered Homes Act 1984) and certain similar properties • registered private hospital (Mental Health (Scotland) Act 1984, section 12) • residential care home (Care Standards Act 2000) • approved bail or probation hostel (Powers of Criminal Courts Act 1973, section 49(1)) • other homes used at least mainly for persons who require personal care by reason of old age, disablement, alcohol or drug dependence or mental disorder, • residential accommodation provided under section 21 of the National Assistance Act 1948 • Abbeyfield Society home
Class B	Religious community	• dwelling inhabited by a religious community whose principal occupation consists of prayer, contemplation, education, the relief of suffering, or any combination of these
Class C	House in multiple occupation (HMO)	• dwelling constructed or adapted for occupation by persons who do not constitute a single household, or • dwelling where the occupier or occupiers each have a contract or licence to occupy only part of the dwelling (see below)
Class D	Resident staff	• dwelling occupied by domestic servant or servants and their family that is occasionally occupied by the owner
Class E	Ministers of religion	• dwelling which is inhabited by a minister of any religious denomination as a residence from which he performs the duties of his office.
Class F	Asylum seekers	• dwelling provided to an asylum seeker section 95 of the Immigration and Asylum Act 1999

Note: as provided by the Council Tax (Liablity of Owners) Regulations 1992 SI 1992 No 551 (as amended)

There have been several cases on composite property and a selection giving insights into the concept is briefly reviewed in Box 6.5.

Valuation

Chapter 16 deals with the valuation of certain types of composite property which have received particular mention in the Council Tax Practice Notes (CTPN), namely:

- dwellings associated with farms (CTPN 2 — second appendix)
- caravans and boats (CTPN 7)

Box 6.5 Cases on the concept of composite property

Office room in a dwelling	*Fotheringham* v *Wood* (VO) [1995] RA 315 Held: that a room with a separate telephone dedicated as an office was a composite property
	Tully v *Jorgensen* (VO) [2003] RA 233 Held: office in a dwelling with equipment supplied by employer (together with personal equipment) of person who had suffered injury and therefore worked from home, was not non-domestic
Guest house / bed and breakfast accommodation	*Skott* v *Pepperell* (VO) [1995] RA 243 Held: a guest house was non-domestic since the usage was more than subsidiary to the occupation as a residence by the tax payer
Non-domestic use	*Mohanti* v *Mackay* [1996] RA 431 Held: where, seemingly, trade as an art gallery was conducted on the ground floor of dwelling, it was non-domestic — the upper accommodation was used as residential

---

- public houses (CTPN2 — first appendix)
- dwellings covered by formula valuations (CTPN 2).

Property straddling boundaries

At the time of the new valuation for council tax a billing authority may have found that some of their dwellings were on the valuation lists of abutting billing authorities and vice versa. It is now provided that, generally, where a dwelling straddles the boundary of two or more billing authorities, it is deemed to be within the area of the billing authority where most of the structure is situated. The provisions and guidance for this include:

- section 1(3) of the 1992 Act
- Part II of the Council Tax (Situations and Valuation of Dwellings) Regulations 1992 SI 1992 No 549 (as amended)
- Practice Note 9: Cross Boundary Dwellings.

The issue is one of finding where the "greater or greatest area" is situated, ie by measuring the superficial area of buildings or structures in each billing authority's area rather than the area of land in the authorities' areas.

Central lists

Certain properties appear on the central rating lists which are used for non-domestic purposes, in effect, national lists of hereditaments (see the Central Rating List Regulations 1989 SI 1989 No 2263). Generally, these are of little interest for council tax purposes but some hereditaments are, seemingly, composite property, ie within them they have domestic property, eg a flat. However, the domestic element of any such property is treated separately because the dwelling is one of the following:

- part of property valued by formula and separated out by apportionment (see p183)
- property capable of being let separately and valued as such.

Exempt dwellings

There are many situations where the owner, developer, occupier or property manager of a dwelling may need to ensure that an exemption from council tax has been given or should be taken up. Thus, in some instances this will be a pro-active step in that they will be able to initiate the release from tax. Box 6.6 shows those dwellings which are exempt from council tax by virtue of the Council Tax (Exempt Dwellings) Order 1992 SI 1992 No 558 (as amended). The box includes a brief indication of the nature of the exemption. Where appropriate, the person concerned should seek to have the property taken out of the tax regime — even for a short period. Also, it may be useful to note future happenings in the caution system or diary so that appropriate steps are taken at the optimum date. For instance, the owner of vacant property is liable for council tax unless the exemption for vacant property applies. Chapter 13 considers exempt property and vacant property in greater detail.

Working at or from home

Many homes are used for working at or from home by the council tax payer or a member of the family who is resident in the dwelling. Issue arises at to whether any part of the accommodation used for this purpose is:

- in *de minimus* non-domestic use and should therefore be ignored, ie the dwelling is wholly within council tax
- in domestic use and non-domestic use but with sufficient domestic use that the non-domestic use is regarded as incidental, ie it is not non-domestic use for rating purposes, ie the dwelling is wholly council tax
- as immediately before but the quantum of domestic use is insufficient to avoid the accommodation being regarded as non-domestic, ie the part accommodation is wholly within non-domestic rating
- the use is wholly non-domestic and there is no domestic use or any domestic use is *de minimus*, ie the part accommodation is wholly non-domestic.

The VOA's Practice Note 8 covers this subject in some detail. Box 6.7 sets out some hypothetical situations of accommodation within a dwelling being used for non-domestic purposes — with attempted solutions (see p77 for dwellings in composite properties and for income tax issues).

Box 6.6 Dwellings which are exempt from council tax

Class	Description
A	Property undergoing structural repair (6 or 12 months)
B	Empty property owned by a charity (up to 6 months)
C	Empty and unfurnished property (6 months)
D	Unoccupied property where the otherwise liable person is being held in detention
E	Unoccupied property where the otherwise liable person is living in a hospital or a care home
F	Unoccupied property where no probate or letters of administration have been granted. (up to 6 months after the grant of probate or letters of administration)
G	Unoccupied property which is prohibited from being occupied by law
H	Unoccupied property which is awaiting occupation by a minister of religion
I	Unoccupied property where the otherwise liable person is living elsewhere to receive care.
J	Unoccupied property where otherwise liable person is living elsewhere to provide care to someone else
K	Unoccupied property where the otherwise liable person is living elsewhere attending college or university
L	Unoccupied property which has been repossessed by a Bank/Building Society
M	Halls of residence
N	Property which is occupied only by students
O	Property owned by the Crown and used as armed forces accommodation
P	Property which is used as visiting armed forces accommodation
Q	Property which comes under the responsibility of a bankrupt's trustee
R	Unoccupied caravan pitches or boat moorings
S	Property occupied only by people under the age of 18 years
T	Unoccupied annexes that are restricted from being let separately by planning controls
U	Property occupied only by people who are severely mentally impaired
V	Property occupied by diplomats
W	Property which is a separate annex and is occupied only by a dependant relative

Box 6.7 Hypothetical non-domestic use of accommodation in a dwelling

Living room	• furnished as a family room • very occasionally used meet work colleagues	• *de minimus* non-domestic use
Spare bedroom	• furnished as living room with computer and filing cabinet • computer used for business and by children • no dedicated business telephone	• business use either *de minimus* or incidental to family use
Garage outside	• office and garage (wholly used as a workshop) both used exclusively for business	• non-domestic use

Property Management, Building and Works

Aim

To show the practical property management aspects of council tax

Objectives

- to identify the treatment of leases and tenancies of property under council tax
- to review rents and council tax
- to describe council tax and houses in multiple occupation
- to establish the concerns which arise with vacant property
- to indicate how to deal with works and other changes to property

Introduction

Apart from home occupiers, many need to be aware of the implications of council tax for the property they are concerned with, including:

- those buying or selling a dwelling or a composite property
- owners of businesses where the property is partly residential (composite property)
- occupiers of the residential part of a business property who are directors or staff
- property managers and facilities managers, eg of investment property
- developers and others having works done to property
- personal representatives.

This chapter examines these implications. It may be useful to note that day to day management of dwellings is undertaken by owner occupiers, landlords, by facilities or property managers or managing agents. Many of the issues and problems are common to all dwellings so some topics dealt with below under a particular section may be of more general or wider interest.

Buying and selling homes

The annual cost of running a home includes the cost of council tax. A prudent buyer of an existing or new house or flat should seek information about the property's present council tax band. In particular, it is important to bear in mind that for a home in Wales the present owner should know the new banding, ie the one from 1 April 2005. (In England under the forthcoming revaluation (now postponed), the banding arrangements were due to change from 1 April 2007 but this is now not so.)

Existing property

Details of council tax payable and the council tax banding should be included in the home information packs (HIP) which must be provided for every owner occupied house or flat put up for sale on the market. Part 5 of the Housing Act 2004 provides for this requirement. The draft HIP Regulations 2006 have been issued by the ODMP for consultation (by 31 December 2005) in which "taxes, levies and charges relating to the property" are specified in schedule 5 paragraph 2(k). (This is due to come into force in 2007.) Until that happens the billing authority will provide details of the council tax for a band and the listing officer will normally supply the following information for existing dwellings:

- the current band
- details of "logged" information concerning alterations to the property which are not reflected in the banding.

It may be noted that not all alterations will have been logged. Also, where a property has been improved by the present owner since the last valuation list came into effect, the improvements will not been reflected in assessment for the current band. On the sale and occupation of the dwelling the listing officer will reassess the property and the result may mean that it will be placed in a higher band.

In due course, information about any grant of planning permission should also be recorded in the HIP. This will be particularly important if the property has planning permission for a non-domestic development. In the meantime enquiries of the local planning authority should result in details being obtained.

New dwellings and conversions etc

The prospective buyer of a newly built dwelling or a unit in a conversion may find that it has not yet been banded. Obviously, the amount of council tax which will become payable is likely to be a budgeting consideration. The listing officer will usually supply an estimate of the band. The applicant for information will find it prudent to supply details of the property, such as the sale particulars, location plan and address, and drawings or room measurements.

Date of leaving the dwelling

Care should be taken to record the dates of vacation and any removal of chattels, supported where possible with documentary evidence, such as:

- copy of any notice to quit
- copy of court orders for possession
- invoices from the removal contractors and others showing work done
- notices for work done, eg cutting off of electricity, gas and water.

Chattels remaining

A conceptual distinction is made between an "unoccupied" property and a "vacant" property. If a taxpayer leaves a property, ie does not reside in it, and takes away all chattels, he or she will not be liable for council tax until any liability begins under vacant property legislation — even then a discount operates. However, leaving the property furnished with chattels will result in continued liability.

Information

Regulations 3 and 4 of the Council Tax (Administration and Enforcement) Regulations 1992 SI 1992 No 613 (as amended) empowers a billing authority to seek information from the following:

- the resident
- the owner
- the managing agent
- other billing and precepting authorities
- electoral registration officers.

This is to enable the billing authority to correctly identify taxpayers. A person, eg a managing agent, who receives a notice should complete it within 21 days or face the prospect of a fine for not doing so (see Chapter 5). The provisions are subject to conditions.

Composite property

A composite property is a building which is essentially a non-domestic property with residential accommodation, eg a schoolkeeper's flat, within it. For the owner, facilities manager or property manager the following may be important for a particular property:

- whether or not it is to be treated as a composite property, eg residential property within a business valued by the formula method (see p183)
- the valuation approach to be adopted (see Chapter 16)
- the payment of council tax — by the employer or the employee
- if the employer pays the council tax, whether or not the employee is treated by the tax inspector as being in receipt of an income tax benefit.

Income tax benefit for "home with work"

Where an employer pays the council tax on the composite property which is occupied by a director, manager or staff, the occupier will normally be treated as being in receipt of an income tax benefit and be charged income tax accordingly. However under the Income Tax legislation exceptions involving occupiers who meet certain conditions, including:

- he or she is occupying because it is necessary for job performance
- he or she will perform their duties better and it is normal business practice to provide accommodation
- he or she is occupying because of a security threat to their person.

Some of the employment positions covered are shown in Box 7.1. They are all in the private, public or voluntary sectors.

Box 7.1 Conditions for exemption from income tax

Private sector	• camping site managers • newspaper shop managers • farm workers • public house managers and certain off licence managers
Public sector	• prison governors and prison officers • police officers • military personnel • diplomatic personnel
Voluntary sector	• ministers of religion and other clergy • sheltered or care housing staff
General	• caretakers

Note: Some of these are subject to conditions in addition to those given above

---

Leased or tenanted property

Information

Normally, the occupying lessee or tenant of a dwelling is liable to pay council tax. A landlord may receive an enquiry from the billing authority as to who occupies a let property. Failure to answer within 21 days may result in a fine. Even if the landlord gives the information about the occupier, liability may arise because the wording of the lease or tenancy requires the landlord to pay.

Tenancy agreement

Normally, the tenant is liable for the council tax and the lease or tenancy agreement should provide for the obligation. Where the landlord is liable, the document should provide this and for any future increases in council tax to be reimbursed to the landlord by the tenant. The rent is, in effect, inclusive of council tax and the tenant pays accordingly.

Rents and council tax

Where the landlord is responsible for paying council tax under the terms of the tenancy agreement, it is a contractual obligation between the parties. However, the billing authority will regard the occupying tenant as liable in the event of the landlord defaulting, ie the tenant is liable despite the tenancy agreement.

Houses in multiple occupation

The Council Tax (Liability for Owners) Regulations 1992 SI 1992 No 551, regulation 2, Class C covers a house in multiple occupation (HMO), ie briefly, one which has, or has been designed for several occupiers who do not constitute a single household or has one or more persons who do not occupy the whole of it. It may be difficult to determine whether it is defined as an HMO. It is important to do so since the owner rather than occupiers will be liable to pay the council tax. (See Chapter 13 for a more detailed account of HMO.)

Management of vacant property

The manager of vacant property will need to "fit" the property into the appropriate category for the payment of or exemption from council tax. Chapters 3 and 10 provide details of any exemption or discount of dwellings and other accommodation. This chapter is, therefore intended to highlight the issues faced by the owner (who may be or may become liable) or property manager. The issues concern the following matters:

- period or periods of emptiness
- alterations to make the dwelling unusable
- demolition of the property
- dates of vacation (see above)
- chattels remaining in the vacant property (see above)
- billing authority's performance targets.

Period of emptiness

In some instances the exemption of a vacant property will exist indefinitely. However, those with up to six months of exemption include:

- completed new buildings
- a dwelling which is unfurnished and vacant
- vacant property following the occupier's death.

Dwellings or other accommodation may be vacant for an indefinite period, including:

- students' accommodation during vacation
- property repossessed by a mortgagee
- property being kept vacate for a member of the clergy
- a dwelling where the occupier is permanently in a hospital or a nursing home
- a dwelling where the owner is detained in prison
- closure due to a closing order or demolition order under the Housing Act 1985, section 267
- property kept closed due to a compulsory purchase order
- property vacant as a result of a planning decision, eg a modification or revocation order.

These cases and others are dealt with in Chapter 13.

Performance targets

One of the performance targets for billing authorities laid down by the Audit Commission is "working with landlords". Although it does not have impact for council tax, it is mainly a social housing and a housing benefit issue. This manifests itself with the following:

- processing times for council tax benefit and housing benefit may encourage landlords to take on benefit claimants as tenants
- generous schemes to pay rent deposits and rent advances on behalf of otherwise homeless tenants to encourage landlords to let to such tenants
- second homes may now only get a 10% discount, which may encourage owners to let them (see Chapter 13).

Proposals

Regulation 5(5) of the Council Tax (Alteration of Lists and Appeals) Regulations 1993 SI 1993 No 290 deals with the circumstances and periods in which proposals are made (see Chapter 15). Some of the entries in Box 7.3 indicate that owners, developers, managers and others may need to deal with a proposal arising, for instance, the implementation of a planning permission. Other opportunities may arise. For example, a person who buys a new home has six months in which to seek a change to the banding. However, if anyone has already challenged the assessment and the proposal was accepted by the valuation tribunal on the same basis, a further appeal on the same basis will not be accepted.

On a revaluation, the new list is published by 31 December to come into operation on the 1 April. An owner-occupier, landlord, tenant (as a taxpayer) or charging authority then has six months from 1 April to 30 November may make a proposal against the banding and so seek to change it. The listing officer may agree the proposal is well founded and make a new entry. If the listing officer does not agree or the altered banding is not acceptable, the person aggrieved may appeal to the valuation tribunal.

Detrimental works

Where the prospect of future occupation is bleak and a liability for council tax has arisen or will arise for a vacant property, it may be possible to alter the property in such a way as to make it uninhabitable. It should then be possible to argue that the actual or prospective liability should be withdrawn. In similar circumstances for liability given in the last subsection, demolition may be the ultimate decision the owner should take. Situations where this is probable include:

- areas where there is no market demand for housing
- a site where the prospect for redevelopment is likely
- cases where there is a planning permission for a site and this will result in redevelopment.

Inappropriate banding

On a revaluation the new list is published by 31 December to come into operation on 1 April, an owner-occupier, landlord/tenant (as taxpayer) or billing authority then has six months from 1 April to 30 November to appeal against the banding and seek a change. Discussion is taking place which may lead to a longer period in which appeals are acceptable after any future revaluation which comes into effect.

Personal representatives

An executor or administrator of the deceased's estate may be required to manage it or oversee the management by a professional property manager. The estate may comprise of:

- owner occupied property
- tenanted property formerly occupied by the deceased as the tenant
- investment dwellings, including composite property
- composite property from which the deceased ran a business.

Generally, the issues and concerns dealt with above may arise during the personal representative's involvement with the estate. Where a professional property manager is employed, most of these matters will be dealt with without the need for the personal representative's direct involvement. However, special concerns arise from the death and the personal representative may need to heed them.

Surviving spouse

The surviving spouse will normally remain in occupation of the family home. If the spouse is the sole occupier after death, the 25% reduction in council tax begins to operate from the date of death. The personal representative should ensure that the billing authority is informed of the changed circumstances. (However, the Registrar of Births, Marriages and Deaths will normally do so soon after the death is registered.)

Owner occupied property

If the deceased was an owner occupier of a dwelling which was a sole or main residence and lived alone, the executor has six months before council tax needs to be paid on the property. The six months runs from the date of the grant of probate or the letters of administration.

One or more second homes which were owner occupied will be charged council tax from the date of death. Of course, if the decision is not already established, the personal representative will need to decide which property was the deceased main residence. It should be borne in mind that capital gains tax should not be an issue, since that tax does not arise on the death of an individual. The choice may be the one which will have the highest liability for council tax in the six months after death.

Tenanted property occupied by the family

The surviving spouse in sole occupation of tenanted property will be eligible for the 25% reduction in council tax from the date of death. The personal representative should inform the billing authority of the death (see above regarding the Registrar of Births, Marriages and Death).

Tenanted property solely occupied by the deceased

Where the deceased was the sole occupier of tenanted property the personal representative may find one of the following circumstances:

- the tenancy comes to an end at the date of death
- the tenancy is a periodic tenancy where notice to quit needs to be served by the tenant or the landlord — the personal representative gives notice (from the date of death) and pays rent
- the tenancy is, for instance, an assured tenancy for a specified period which will come to an end by the passing of time.

In these cases the period of exemption will last until the tenancy comes to an end or is disposed of by the personal representative. The payment of rent by the latter is not eligible for housing benefit.

Investment property

The occupying tenants will normally be liable for council tax but where the landlord (the deceased) let the property on an inclusive basis, the personal representatives assume responsibility for the payment of the tax on each property until the property is distributed. Each person who benefits from the distribution will be liable from the date on which the property is received.

Works etc to property

For dwellings, changes of use, new buildings, improvements and alterations may have implications for council tax payers which should be dealt with speedily. Box 7.2 highlights some of the practical points which may need to be addressed, including:

- planning permission
- the condition of the property, ie repair or inherent defect
- various kinds of works, eg to meet the needs of a disabled person
- proposals to change an entry in the list (see also Chapters 15 and 16)
- new buildings and completion notices.

Box 7.2 Council tax issues arising when works are carried out to property

Grant of or the existence of a valid planning permission	• may result in a proposal to alter the valuation list • planning permission for a material change of use of a dwelling to some non-domestic use is likely to result in the dwelling being removed from the valuation list • rating list being altered with the inclusion of the property as a non-domestic hereditament
Condition of the property	• when property is in poor state of repair the taxpayer may feel justified in making a proposal to have the assessment reduced • condition is assumed to be reasonable for council tax valuations (see regulation 6(2)(e) of SI 1992 No 550) • state of repair should not make a difference to the value for the tax
Inherent defect	• an inherent defect in a property may be such that works are frequently needed to make good damage caused to property • the defect, by definition, cannot be remedied or could only be remedied at a cost which is considered unreasonable • appropriate, therefore, to make a proposal to have the value reduced
Construction of new dwellings	• check for exemption or relief — may depend on the each occupier's status • likely need to negotiate with the listing officer on the assessments • prospective buyers may want an estimate of the banding • a liability arises when a chargeable dwelling becomes occupied or if vacant, on expiry of a period of six months after the property became substantially complete • procedure for completion notices to be observed • occupation by the builder of any site office or sales office or both may be liable to business rates
Self-build by the site owner	• living on the site in a caravan, for example, likely to result in its assessment to council tax • living in the partially completed new dwelling, needs a certificate of habitation from the building control officer • similarly, habitation may result in assessment and liability for council tax
Alterations to a building	• any change in capital value may result in possible change in prospective banding • improvements (result in increase in value) only result in a possible step up of banding on the sale of the property or on national revaluations which are now likely to occur by order (the government does not want to discourage improvements • "negative" improvements may result in a step down in banding — make a proposal to the VOA
Demolition of a building	• when the building is first vacated seek cessation of payments to tax • partial demolition may prevent incidence of tax under vacant property charge or at least reduce the value of the property • after demolition make a proposal to remove from the list

Box 7.2 continued

Works of repair and maintenance	• repair and maintenance may result in an increase in value • exceptional standards of repair and maintenance, so that the dwelling is outstanding, may increase the value over others (a possible revaluation issue but unlikely) (see above)
Adaptations for a person with a physical disability	• works to a dwelling which meets the needs of a person with one or more disabilities • they are ignored in an assessment for council tax • prudent owner or manager may need to record the works and why they were installed (particularly to show the situation where the disabled person no longer resides in the dwelling) • see regulations 6(2)(g) and 6(4) of SI 1992 No 550, as amended
Conversion of a single dwelling into flats	• may not affect the banding • report to the VOA; each self-contained flat will give rise to a separate council tax liability • separate assessments for each flat may mean that the burden for the owner (retaining a flat) will be reduced
Conversion to flats of the upper floors of a shop or offices	• the flats will become chargeable dwellings • business rating (and an composite property element) will, no doubt, change • planning permission will normally be required • capital allowances may be available for the cost of the works
Adaptations (probably minor) for rent a room income tax relief	• should not result in any change to the banding status • single person discount (if being claimed) will end when room is let (unless to a student, or person under 18 or lodger in receipt of income support or job seeker's allowance
Conversion of a dwelling to a holiday cottage etc (or a change of use)	• probably takes a hitherto chargeable dwelling into business rating • business comes within the income taxation regime, eg possible claim for capital allowances
Change of use to non-domestic	• make proposal to remove from the valuation list

Completion notice

Section 17 of the 1992 Act and schedule 4A to the Local Government Finance Act 1988 provide the procedure for completion notices to apply to dwellings. For a newly constructed dwellings or a major reconstruction, a completion notice is issued by the billing authority to the owner. The notice states the date on which the billing authority reasonably expect the building to be completed. The notice's date is the trigger the billing of council tax which becomes payable three months later in the event of the building being unoccupied (the tax will be at the 50% discount). Of course, if the dwelling is occupied before the three months is expired the liability commences from the day of occupation. (Chapter 18 gives details of an appeal against a completion notice (see p205).)

Part 4

Administration and Recovery of Council Tax

Administration of Council Tax

8

Aim

To describe the billing authority's functions of determining, billing, collecting and enforcing the payment of council tax

Objectives

- **to show a schematic structure for a typical billing authority's revenue office**
- **to show how the bills are dispatched and the money collected**
- **to tentatively outline some concerns about the process**
- **to describe the ways of enforcing payment**

Introduction

In Chapter 1 the billing authorities are listed in Box 1.7. They have functions which include:

- identifying taxpayers
- billing the taxpayers as appropriate
- collecting the council tax
- enforcing payment of the council tax
- administering council tax benefit
- countering fraud and corruption by taxpayers, claimants and others
- promoting the revenue services (see Chapter 5).

This chapter explains the determination of the council tax, billing, collecting tax and enforcing payment. Issues of council tax benefit, counter-fraud measures and performance are covered in Chapters 9, 10 and 11, respectively. Disputes and complaints are dealt with in Chapters 17 and 18.

Structure and organisation of a revenue service

A revenue service will normally deal with not only council tax and non-domestic rating but also housing benefit and council tax benefit. Assuming the billing authority's area contains say, 40,000 dwellings, the total revenue staff will be up to about 50 persons divided into three departments, namely:

- local taxation dealing with billing and enforcement for both council tax and non-domestic rating (business rates)
- housing benefit and council tax benefit
- cash office.

Local taxation may have one of several configurations — typically it will be divided into three functions or sections dealing with the following:

- council tax determination of liability, billing, reminders, final notices and direct debit collection
- non-domestic rates
- enforcement of council tax and business rates, recovery by summons and subsequent remedies.

Housing benefit and council tax benefit is likely to include the following functions:

- assessment
- welfare and visits to vulnerable families or individuals
- fraud investigation
- recovery of overpaid housing benefit and council tax benefit.

There are many variations on the above revenue service structure. For example, a number of authorities have attempted degrees of generic working with limited success. These range from the moderate approach of combining billing with enforcement to the extreme examples of combining benefits with billing.

Obviously, the scale of operations of a revenue office will depend on such matters as:

- the number of dwellings in the billing authority's area
- similarly, the number of non-domestic properties
- the level of claimants and recipients of council tax benefit.

However, where part of the work is outsourced, eg to civil enforcement agencies in the private sector, the internal work should be less.

Outsourcing

A billing authority may contract out most aspects of its work on administering council tax, eg billing and enforcement, to companies in the private sector. The power to do so is given by the Local Authorities (Contracting Out of Tax Billing, Collection and Enforcement Functions) Order 1996 SI 1996 No 1980. The billing authority should consider a requirement that the contractor has appropriate insurances, eg professional liability insurance, and will adhere to codes and standards.

Determination of council tax

Under sections 30 to 37 of the 1992 Act the tax is determined by the billing authorities. It is then billed to the occupiers of chargeable dwellings or, if appropriate, the owners. The "unit" amount of council tax having had regard to:

- its estimated spending on statutory duties and discretionary opportunities
- government grants and the like
- the total number of adult residents
- the number of adult residents excluded from the total
- amount of any reduction caused by means tested council tax benefit
- the amounts sought by the precepting authorities
- estimates of other receipts and obligations, eg interest on loans.

Precepting authorities, eg county councils, police authorities, town councils and parish councils inform the billing authority of their council tax requirement and receive a share of the council tax as a precept.

Billing and collection

Once the taxpayer is identified (see p57), he or she is billed for the amount due; which must be for the exact amount. The amount is set in accord with sections 30 to 37 of the 1992 Act. The amount collected in respect of each dwelling must have regard to the following:

- the level of funds needed by the billing authority for the year
- the valuation band for the property
- transitional adjustments following revaluation
- any exemption available in respect of the property
- the number of people in the dwelling
- any discounts, reductions or reliefs available to the occupier (see Chapter 13)
- council tax benefit.

Until it is proved otherwise the billing authority will bill and enforce collection against the individual who has been identified as such (see *Des Salles d'Epinox* v *Royal Borough of Kensington & Chelsea* [1970] 1 All ER 18).

Restrictions

The government may restrict or cap the amount which a particular billing authority wishes to impose (see p12).

Payment of council tax

The Council Tax (Administration and Enforcement) Regulations 1992 SI 1992 No 613, as amended, provide the ways in which payment of council tax is made and where necessary enforced. The billing authority will normally enforce payment in the event of non-compliance.

Payment schedules

Council tax may be paid by one of three statutory schedules of schedule 1 to SI 1992 No 613 — from which the taxpayer is entitled to choose. However, a taxpayer and the authority can agree a different schedule between them if they so desire. The three provided schedules are

- over 10 monthly instalments starting in April, (usually the first day of April) of the financial year that gives rise to the tax, an ending with a tenth payment on 1 January
- half yearly in advance, payable in April and October; usually the first day of the month
- annually in advance, payable in April — usually the first day of April.

Payment methods

Most authorities accept a wide variety of payment methods including cash, cheque, payment over bank counter or post officer, debit card, credit card, standing order, internet, and direct debit. Electronic demand, eg e-mail, may be made by virtue of the Council Tax and Non-domestic Rating (Electronic Communications) (England) Order 2003 SI 2003 No 1234, as amended.

Discounts for certain types of payment

Authorities have power, if they so desire, to offer a discount on council tax for paying a lump sum annually in advance or for paying by direct debit (see regulations 25 and 26 of SI 1992 No 613). If offered, these particular discounts must be given to all taxpayers who fit the criteria, not just those converting to lump sum payment or payment by direct debit.

Back-dating payments

For a variety of reasons, back-dating of liability may arise. Examples include:

- when the authority discovers the identity of a liable party, which can be quite late especially where the party has evaded identification
- when a liability in respect of a dwelling originally listed as a single dwelling but later found to be a house in multiple occupation.

The *Valuation Tribunal Service LPAC Newsletter*, October 2004 gave an example of the latter in a case before the West Midlands (West) Valuation Tribunal. In the event the billing authority was unsuccessful in the extent of back-dating — wanting to go back to December 1999. It was found that although the billing authority had records which would have enabled it to take action in that year, it failed to issue the bill as soon as reasonably practicable and therefore was not permitted to backdate that far — it pressed the Valuation Tribunal for 1999 but the Valuation Tribunal decided April 2003 would be fairer to the taxpayer.

Over-payment

From time to time a taxpayer may pay more council tax than should be paid.

Typical examples are when:

- a liable occupant moves to a new residence having paid council tax in advance
- an award of council tax benefit on a part paid account gives rise to a credit.

The legislation provides for the billing authority to remit the amount of excess. If the billing authority refunds too much or refunds in error, then recovery of the erroneous refund cannot be enforced by applying for a council tax liability order (see *AEM (Avon) Ltd* v *Bristol City Council* [1998] RA 89). It would have to be enforced, if necessary, through the county court.

Non-payment

If the taxpayer fails to pay a council tax instalment, a reminder is sent by the billing authority. If the reminder is not complied with, the billing authority will apply to the court to issue a summons and a liability order.

Detail of the process

If an instalment (usually one of 10 monthly instalments) is late by a day or more, the authority may issue a statutory reminder notice. The reminder notice has a duel effect. First, it requires that the instalment or instalments that are in arrear be paid within seven days of the date of issue of the notice. Second, it dictates that if the arrear is not paid within seven days, then the remainder of the whole year's council tax becomes payable in full in advance, and must be paid before the end of the 14th day after the date of issue of the notice.

If the authority has received neither the arrear within the first seven days, or the remainder of the full year's council tax by the 14th day then it may apply to the magistrate's court for a liability order to enforce collection of the full year's council tax. In practice a billing authority may permit a few days longer than this for compliance before applying to the court for a liability order.

There are also notices known as "second reminders" and "final notices". The names of these are somewhat misleading, as they only come into play where an initial reminder notice is complied with but the taxpayer is subsequently late with another instalment later on in the year.

Enforcing payment

Part VI of SI 1992 No 613, as amended, provides for the enforcement of council tax. Thus, billing authorities enforce the collection of unpaid council tax principally by applying to the magistrate's court for a liability order against the defaulting taxpayer. A liability order empowers the billing authority to proceed with any of the following enforcement remedies:

- demand employment and other income details
- deduction from job seeker's allowance, income support or pension credits
- attachment to earnings
- petitioning for the debtor's bankruptcy
- distress procedures by "civil enforcement agent" (bailiff action)

- attachment to a defaulting councillor's allowances (if the debtor is a local councillor)
- application to the county court for a charging order to register a charge against the debtor's property
- application to the magistrate for committal to prison following unsuccessful bailiff action
- rarely, the council may deduct directly from other types of benefit to recover council tax benefit that has been overpaid because of the claimant's misrepresentation of failure to disclose a material fact.

The procedures, evidence and defences for an application for a liability order are discussed in Chapter 18.

Demand for income and employment details

Although a demand for income and employment details may not in itself seem a remedy, it does in fact result in quick payment in full in very many cases. It is used mainly to obtain information to impose an attachment of earnings order.

Thus, after a liability order has been granted the billing authority will usually write to the debtor demanding income details which must be provided to the authority within 14 days (see regulation 36 of SI 1992 No 613). Details required include:

- employer's name and address
- details of source of income support or jobseeker's allowance.

Failure to provide this information is a criminal offence punishably through the magistrate's court with a fine (at the time of writing is up to £500). The authority will use the information to make attachments to earnings or direct deductions from income support or jobseeker's allowance where possible. If this information is already know to the authority this action may be taken immediately without sending a demand for income information.

Deductions from certain benefits and pension credits

The Council Tax (Deductions from Income Support) Regulation 1993 SI 1993 No 494 provides for deductions from certain benefits. While persons on income support (IS) or jobseeker's allowance (JSA) usually receive 100% council tax benefit, they may have unpaid council tax liability from a period before they began to receive those benefits. If this liability remains unpaid then the billing authority, after a liability order has been granted, may apply to the Department for Works and Pensions for direct deductions to be made from any IS or JSA being paid to the debtor. At the time of writing the current rate of deduction imposed is £2.85 per week. Councils are also empowered by paragraphs 6 and 12 of schedule 4 to the Local Government Finance Act 1992 as inserted by the State Pension Credit Act 2002 to make similar deductions directly from pension credits, both guarantee credit and savings credit aspects.

Attachment of earnings

After a liability order is granted, the billing authority may make a council tax attachment to earnings order (CTAEO) (see regulations 39 to 43 of and schedule 4 to SI 1992 No 613). This is done without further recourse to the court. The CTAEO is directed to the debtor's employer who is obliged to comply with it by deducting sums from the debtor's remuneration and paying it over to the billing authority.

Amount to be deducted

The amount to be deducted is based on a percentage of income which varies according to income bracket. The percentages are also varied from time to time by statute to reflect inflation. The appropriate percentage deduction is applied to the pay after having deducted tax and national insurance. Child tax credits, sick pay and redundancy pay are not attachable as earnings. Up to two CTAEOs can be imposed simultaneously, the second order being immediately applied to the residue of earnings after having deducted the first order. At the time of writing the percentage deductions for each income bracket are as shown in Box 8.1 below. There are similar tables for daily and weekly paid persons that give rise to similar proportions of pay being deducted. It is a criminal offence for the employer to fail to apply the order if the subject is in his or her employment, punishable with a scale 3 fine (currently up to £500) through the magistrate's court.

Box 8.1 Council tax — attachment of earnings: monthly rate

Net earnings	Deduction rate (%)
Not exceeding £220	0
Exceeding £220 but not exceeding £400	3
Exceeding £400 but not exceeding £540	5
Exceeding £540 but not exceeding £660	7
Exceeding £660 but not exceeding £1040	12
Exceeding £1040 but not exceeding £1480	17
Exceeding £1480	17 in respect of the first £1480 and 50% in respect of the remainder

---

Bankruptcy

A billing authority may petition the High Court (for London) or its county court for the debtor's bankruptcy. In fact, where the debtor owes more than £750, it seems that billing authorities are increasingly using powers under Part IX of the Insolvency Act 1986 for bankruptcy proceedings as an enforcement remedy. The use of the bankruptcy remedy may be preferable to seeking the debtor's committal to prison for the reasons given in the Court of Appeal decision in *Griffin* v *Wakefield Metropolitan Borough Council* [2000] RVR 226, namely:

- bankruptcy is a legitimate remedy
- any remedy given by statute is legitimate
- bankruptcy is not inherently more draconian than commital.

The judge also referred to the following:

- prison is a costly remedy
- the debt is not always recovered by this means of this remedy alone
- the Human Rights Act 1998 may be invoked against committal to prison.

Distress or bailiff action

The enactment for distress or bailiff action is SI 1992 No 613, as amended, regulations 45, 45A, 46 and schedule 5. The prerequisites for bailiff action are:

- a liability order must have been granted
- the liability order is at least partly unpaid
- the liability order has not been quashed or otherwise rendered invalid through a court of law
- the billing authority must have given seven days notice in writing of both its intention to instruct a bailiff, and details of the schedule of allowable bailiffs fees and charges.

Following the making of a liability order by the magistrate's court, many authorities will send a demand for employment details and a notice of bailiff action including bailiff fees, simultaneously on the same document. Any valid liability order meeting the above requirements may be acted upon by way of bailiff action. In practice it is usually those liability orders for which suitable opportunities for attachment to earnings or benefits have not been available, that are passed for bailiff action. The bailiff may be directly employed by the billing authority, or more often, will be a company contracted by the billing authority for the purpose of executing liability orders by distress. Any bailiff executing a council tax liability order must be suitably certificated by the county court to act as such.

At the time of writing the schedule of permissible bailiff fees for council tax liability is summarised in Box 8.2.

Remuneration of contracted bailiff companies

Due to the competitive nature of the industry it is not uncommon for bailiff companies to be contracted on the basis that they may retain the fees and charges they collect under the provisions in the above table, without further remuneration.

A debate has arisen about the reasonableness of bailiff companies profiting from permissible fees that they can impose on the debtor. This is considered acceptable, subject to any successful future challenge, on the basis that the first and second visit fee prescribed by the statutory schedule of fees are not linked to costs actually incurred. Also the billing authority is under a duty to achieve best value, which means appointing the least expensive bailiff who can satisfy the specific quality, conduct, and collection effectiveness requirements of the contract.

Box 8.2 Permissible bailiff's fees for council tax liability

First visit — attempting to levy but no levy is made	£22.50
Second visit — attempting to levy but no levy is made	£16.50
For levying distress the lesser of	Reasonable costs and fees or 22.5%. on the first £100 of the sum due, 4% on the next £400, 2.5% on the next £1,500, 1% on the next £8,000 and 0.25% on any additional sum;
Visit — with a view to removing goods (where goods are not removed)	Reasonable costs, fees, expenses incurred
Removal and storage of goods for sale	
Valuing an item at the debtors request	
Closed possession of goods	£14 per day
Walking possession of goods	£11
Auctioneer's fee	Up to 15% plus any reasonable costs of advertising

Goods protected from distress

Regulation 45 of the Council Tax (Administration and Enforcement) Regulations 1992 SI 1992 No 613 provides that items which cannot be subject to distress include items of equipment necessary for personal use by the debtor in their employment, business or vocation. These items cover tools, books, and vehicles so used. Items which are necessary for the satisfaction of basic domestic needs of the debtor and family are also protected. These necessities include clothing, bedding, furniture and necessary household equipment.

Other things that are protected under common law and statute include:

- goods which are not the property of someone named on the liability order
- property of ambassadors, diplomats, and other international organisations — protected by Diplomatic Privileges Act 1964
- goods which are possessed by a landlord who has levied for unpaid rent
- items possessed by the HM Revenue & Customs — protected by the Taxes Management Act 1970 (as amended)
- Crown property
- items in use (such items as tools are privileged while they are actually in use)
- money (though it is distrainable if it is in a closed container of some sort)
- goods in the mail so protected by section 64 of the Post Office Act 1969
- goods which have been seized into the custody of the law
- fixtures on the property
- goods for which a contract for sale has been agreed, but the transfer of ownership of goods is (in accordance with the contract) not due to take place until the full purchase price is paid

- perishable goods which cannot be retained in the same condition prior to distraint
- gas fittings so protected by section 27 of the Gas Act 1972
- electrical fittings so protected by section 57 of the Electricity Act 1947
- good on hire purchase (there are some exceptions) — protected by sections 87 to 90 of the Consumer Credit Act 1974

There are other items which are protected but they are not likely to arise in a council tax situation, eg railway carriages — under the Railway Rolling Stock Protection Act 1872.

Issues regarding distress for council tax

The following issues regarding distress are emphasised:

- a bailiff can only distrain on goods that belong to the debtor in question
- a bailiff must gain access to the goods which are to be levied — he or she may not therefore impose a levy fee based on what goods is thought may be in a dwelling, or goods that can be seen from outside the dwelling (see *Evans v South Ribble Borough Council* [1992] 2 All ER 695)
- a visit fee and a levy fee can not be imposed on the same visit. The visit fee is only applicable if there is no successful levy (see schedule 5 to SI 1992 No 613)
- once the bailiff has started removing possessions, he may continue until he has finished before accepting any payment and may include the costs associated with removing the possessions
- forcible access to a home is theoretically possible under some very extreme circumstances where the bailiff is in walking possession of goods and there is a repeated lack of cooperation and failure to keep appointments by the debtor for the bailiff to collect the goods
- forcible entry is unlikely to be carried out for council tax, as this form of action is very vulnerable to successful legal challenge
- where there are more suitably valued items reasonably available for distraint, the bailiff can not take something that will realise a great deal more than the sum to be collected (see *Steel Linings Ltd v Bibby & Co* [1993] RA 27).

Effectiveness of bailiff action

There is a commonly held view that bailiff action ought not to be used in modern society because it is thought to be "ineffective" and not needed. In fact bailiff action recovers by far the majority of council tax subject to liability orders. For most authorities this amounts to tens of thousands of pounds per month. Most of this is collected under threat of removal of possessions without an actual removal taking place. An authority with say 50,000 residences will typically secure anywhere between 2000 and 6000 liability orders for unpaid council tax each year, and typically more than half of these will be administered by bailiffs. Without this remedy the enforcement system would be largely ineffective, non-payment would grow, and the lower collection rate would lead to higher council tax bills.

Insufficient means to pay

Of course, bailiff action is not taken against those found by statutory assessment to have insufficient means to pay council tax — such persons are awarded council tax benefit accordingly (see Chapter 9).

Bailiff action or attachment to earnings

While the more modern remedies like attachment to earnings have slightly reduced the amount of bailiff action necessary, this is by no means a comprehensive alternative. For example, attachment to earnings is not effective against the self employed.

Scotland

In Scotland, section 16 of the Debtors (Scotland) Act 1987 provides a more limited list of items for enforcement purposes — it replaces distress under a multi-sourced law of protection from distress for goods which existed prior to the Act coming into force.

(However, a debt arrangement scheme (DAS) may be available to a defaulting council tax payer under the Debt Arrangement and Attachment (Scotland) Act 2002. Here, the debtor with two or more debts agrees a debt payment programme (DPP) which must be approved by the debt administrator.)

Attachment of a councillor's allowances

Regulation 44 of SI 1992 No 613 provides that after a liability order has been granted, the billing authority may make an attachment of allowances if the subject of the liability order is an elected member of a local or county council. The deduction will continue at a rate of 40% of the councillor's allowances until the liability is paid.

Charging order

Regulations 50 and 51 of SI 1992 No 613 provide for charging orders. Thus, where a debtor owns property that gave rise to the unpaid council tax, the billing authority may under the power of a liability order, seek from the county court a charging order against the property. The property then becomes security for the debt, which will be realised when the property is sold. The minimum debt for this action is £1000.

Committal to prison

Where the bailiff is unsuccessful in recovering the unpaid council tax, the billing authority may apply for the defaulter to be committed to prison for up to 90 days for each council tax liability order. The statutory authority for this is regulations 47 and 48 of and schedule 2 to SI 1992 No 613.

The defaulters are summoned to the magistrate's court. Most culpable defendants who have evaded action are sensible enough at this stage, to settle the unpaid council tax and all the accumulated council's costs, and bailiff's costs. If they neither pay in full nor attend the court, they are arrested and ultimately forcibly brought before the court, to give account of their conduct in non payment of council tax. If on hearing the evidence presented by the council and the defaulter, the magistrates decide the defaulter has culpably neglected, or wilfully refused the pay, the magistrate may impose a prison sentence of up to 90 days for each council tax liability order. A prison sentence may be made, but postponed on payment terms. If the magistrate finds there is no culpable neglect or wilful refusal they may write-off part or all of the unpaid council tax. The magistrates may also decline to decide and dismiss the case, usually with a recommendation made that the defendant begin paying a recommended amount, lest the council re-apply for their committal to prison.

(Chapter 18 gives a brief account of the committal procedure before the magistrate's court and some aspects of case law.)

Effectiveness of committal action

It does not appear that a comprehensive national study of effectiveness has been carried out. One study involving a sample from two district billing authorities showed that 50% of the sums for which committal applications were made were fully paid within six months of the applications. The majority of the sums that remained unpaid were for defaulters who absconded without trace. Another local study showed that 98% of those actually committed to prison managed to raise sufficient money in the first 72 hours of their imprisonment, to pay the unpaid sum, and thereby be released from prison.

The power to imprison must first be used in its coercive nature, ie by postponing the imprisonment on payment terms, before it is executed in its punitive nature, ie by committing to prison. If a prison sentence is served in full, the council may still accept payment if it is offered. This is because there is no legal obligation to write off the debt. However, if no offer is made, the billing authority cannot take further enforcement action of any kind to collect it; so in practice it is usually written off on the basis that it is irrecoverable.

Fraud

A written policy on anti-fraud and anti-corruption is a performance indicator on benefits, including council tax benefit, required of billing authorities (see p112). Chapter 10 examines counter-fraud and counter-corruption policy and practice in some detail.

CAB involvement

The local citizens' advice bureau provides information and help to taxpayers and council tax benefit claimants. Some offices have staff who are specially qualified to assist claimants or potential claimants who have problems or who wish to make claims.

At the national level, the National Association of Citizens' Advice Bureaux has a comprehensive website on council tax and council tax benefit. The NACAB or an individual bureau is often a source of feedback to billing authorities and others on the issues the bureau clients face.

Council Tax Benefit

Aim
To examine the nature of and eligibility for council tax benefit

Objectives
- **to describe the main types of council tax benefit**
- **to identify eligible claimants**
- **to describe how a claim may be made**
- **to show with examples how council tax benefit may be calculated**

Introduction

From 1993 financial support in the form of council tax benefit towards the cost of council tax has been available to eligible households on a low income. It is administered by local authorities alongside housing benefit. It is means tested and can be awarded up to 100% of council tax liability. Persons on income support or job seeker's allowance are usually entitled to 100% council tax benefit. The council tax benefit scheme is covered by the following principal enactments:

- section 130 of and schedule 9 to the Local Government Finance Act 1992 (the 1992 Act) which embeds council tax into the social security legislation
- section 131 of the Social Security Contributions and Benefits Act 1992
- the Council Tax Benefit (General) Regulations 1992 SI 1992 No 1814, as amended
- Disability Discrimination Acts 1995 and 2005.

They provide for two forms of council tax benefit, namely:

- the main council tax benefit
- colloquially, the second adult rebate (or formally, "alternative maximum council tax benefit").

Numerous statutory instruments have been introduced after 1992. They govern the way in which council tax benefit is applied, particularly in relation to tax credit and pension credit.

This chapter describes the role of the claimant and the billing authority in council tax benefit, ie it covers claims and administration. Counter-fraud administration of benefits is dealt with in Chapter 10.

Claimants

Any payment of council tax benefit follows a claim being made by an eligible individual (see regulation 61 of SI 1992 No 1814). It is submitted to the billing authority responsible for collecting the council tax.

Who can claim?

A claim for council tax benefit may be made by any person who is liable for council tax. If such a person is unable to claim due to physical or mental incapacity, then a claim can be made on his or her behalf by an appointee. An appointee is a person of 18 years or more who has successfully applied in writing to the local authority to represent the incapacitated claimant. A person holding enduring power of attorney (EPA) or an official receiver may also act for such a claimant. (The Mental Capacity Act 2005 will introduce the "lasting power of attorney" in 2007 — it will replace the EPA, although exant EPAs at that time continue.) In Scotland the Adults with Incapacity (Scotland) Act 2005 covers forms of power of attorney.

Couples

Only one claim may be made by a couple, ie either party may claim for the two of them — they decide. A couple is one where two persons:

- are married
- live together as man and wife
- are in a civil partnership within the meaning provided by the Civil Partnership Act 2004 as it applies to same sex couples.

In the event of any uncertainty on the part of the person or the local authority, it is the local authority that will decide if the persons are a couple for the purposes of claiming and calculating council tax benefit. It is possible for a couple to live mainly in separate dwellings and still be a couple for the purpose of council tax benefit. For example, if one party mainly lives elsewhere for the purpose of their employment. (The provisions have implications for the administration of council tax as provided for by enactments, for example:

- the Council Tax (Civil Partners) (England) Regulations 2005 SI 2005 No 2866
- the Council Tax (Exempt Dwellings) (Amendment) (England) Order 2005 SI 2005 No 2865.)

Joint tenants who are not couples

Where joint tenants are not couples, it will be assumed for the purpose of calculating benefit, that each liable party is responsible for his or her proportion of the council tax. Note, however, that the responsibility to pay the remainder of the council tax, after benefit is calculated and awarded, usually falls jointly and severally to all the liable parties (joint and several liability is discussed in Chapter 12 (see p134)).

A billing authority is sometimes required to determine whether a couple are merely sharing a dwelling or living as partners as this can considerably affect their benefit entitlement. Several cases, eg *Crake & Butterworth* v *Supplementary Benefit Commission* [1982] 1 All ER 498, address this issue. The criteria to be considered include:

- whether they share the same household
- the stability of their relationship
- the financial arrangements
- where a child lives in the dwelling, the extent of any shared responsibility
- public acknowledgement that they are a couple.

Also, if the Department for Works and Pensions (DWP) has awarded pension credit, income support or job seeker's allowance as a couple, the billing authority should accept this (see *R* v *Penwith DC HBRB, ex parte Menear* [1991] 24 HLR 115. However, this is not the case where the claim to the DWP was based on fraud (see *R* v *South Ribble Borough Council HBRB, ex parte Hamilton* [1999] QBCOF 99/1021/4.

Capital

Anyone who has more than £16,000 of assessed capital is not entitled to council tax benefit. The first £3000 of a person's estate is excluded from the total of assessed capital, ie it does not affect entitlement to benefit. (It is proposed to increase the £3000 slice to £6000 in April 2006.) Capital includes many types of asset, including such things as:

- savings, ie premium bonds, cash, bank or building society balances
- tax refunds
- investments, such as stocks and shares
- property other than that in which the claimant lives.

Some things are not considered capital for the purpose of calculating council tax benefit, including:

- the home in which they live and from where they are applying for council tax benefit
- any capital held by a person receiving income support or job seeker's allowance
- personal assets such as a car or caravan
- the surrender value of a life insurance policy
- some types of compensation payments
- some types of assets held in trust.

Any back dated lump sum payments of other benefits are usually disregarded as capital for a period of 52 weeks.

In considering jointly held capital, it should be assumed that the joint owners of the capital each own an equal share unless an alternative share of ownership is distinct and known (see *Secretary of State Work and Pensions* v *Hourigan* [2002] EWCA Civ 1980).

Once the capital has been assessed, the claimant is treated as having £1 per week income for each £250 capital they have in excess of £3000. The resultant income figure is known as the "tariff income" and is added into the claimant's total income for the purposes of calculating their entitlement to council tax benefit.

Equity release

Owner occupiers who take the opportunity to release capital from their homes may find that their entitlement to council tax benefit is affected. Any increase in "liquid" capital over the statutory minimum for the calculation of council tax benefit will result in a loss of the benefit (see regulations 28 to 37 of SI 1992 No 1814).

Making a claim

Claims for council tax benefit are in writing, usually on a comprehensive composite form developed and approved by the authority for the purpose of assessing both council tax benefit and housing benefit (see regulation 62 of SI 1992 No 1814). The form is likely to include a statement to the effect that a claimant can be prosecuted for failing to supply the correct information (see Chapter 12). There are some circumstances where less information is required when some authorities may use a shortened version of the form. For example, where the application is for a claim to be continued and confirmation is needed that the claimant's circumstances have not changed since the last claim. Technically a letter could constitute a valid application if it contained enough information, but in practice this is seldom, if ever, approved by a billing authority.

In the future it is likely to be possible for claims to be made by telephone, and a confirmation statement issued to the claimant to provide their signature. At the time of writing 56 billing authorities are piloting such a scheme in conjunction with the DWP and the Employment Service. The scheme is known as the "ONE Service" and is designed to enable applications for many benefits including council tax benefit to be made with a single telephone call. When an income support or job seeker's allowance claim is made to the DWP, the DWP will usually forward the details to the billing authority for council tax benefit to be assessed.

Households that include students

Full-time students do not normally pay council tax, particularly those living in a hall of residence or in "student accommodation" (see regulations 38 to 50 of SI 1992 No 1814). Some students are required to pay council tax and will not usually be eligible for council tax benefit, but there are exceptions, including:

- a student with dependent children
- a student who is disabled
- the partner of a student who pays council tax.

However, some students may be able to claim second adult rebate (see below). Students may sometimes be entitled to other discounts or exemptions (see Chapter 13).

Effective date of claim

A claim is usually effective as of the day it is received by the local authority's designated office or the DWP. However, if the claimant is in receipt of income support or income based job seeker's allowance, then the effective date of the council tax benefit claim is the first day of entitlement to income support, or the day on which application for job seeker's allowance was made. If a submitted application is not supported by all the required evidence it will still be considered to start on the date of its submission so long as the required evidence is provided within four weeks. The authority can extend the four week period if it deems an extension reasonable for the particular circumstances.

Backdating the claim

Most council tax benefit application forms include a page on which a request for backdating can be made (see regulation 62(16) of SI 1992 No 1814). Council tax benefit can sometimes be backdated up to a maximum of 52 weeks from the making of a claim; however the circumstances need to be somewhat exceptional. To be successful the claimant must show that there were good reasons for failing to apply earlier. They must also show that the reasons were continuously valid throughout the period in which a claim could legally have been made earlier. (This is known as the need to show "continuous good cause" for not having applied for council tax benefit earlier.) The power to determine whether there has been good cause lies with the local authority. The burden to prove it lies with the claimant. Each case is to be considered on its merits whereby a test of reasonableness is applied, ie would the facts have lead a reasonable person to fail to claim benefit at the proper time? A claimant needs to demonstrate that they did everything which a reasonable person in their particular circumstances would have done.

Typical facts considered include:

- age
- health
- background
- knowledge of social security system
- information previously provided to the claimant
- information the claimant could reasonably have obtained.

There is considerable scope for varied interpretation; however Box 9.1 gives a summary of the type of reasons and rationale applied.

Special circumstances of claimants

War disablement pension and war widow's pension can be excluded wholly or partly for all such cases at the local authority's discretion. Similarly, claimants on income support or jobseeker's allowances are considered to have no income and will be awarded 100% council tax benefit.

Box 9.1 Typical reasons and rationale for the backdating of council tax benefit

Reasons in back-dating requests	Assumption	Reasonableness tests for awarding backdate
Ignorance	It is normally reasonable to expect a liable party to be aware that council tax benefit exists and and may be entitled, especially as it is mentioned on bills and reminder notices etc. The claimant must show reasonable causes for his or her ignorance	Do the facts indicate it was reasonable that such a claimant would be ignorant of council tax benefit, eg illiteracy? Was it reasonable for the claimant to believe there was no benefit entitlement, eg wrongly advised? Did the claimant have a reasonably held firm belief that being self employed he or she could not claim
Failure to enquire	It is normally reasonable to expect a liable party to inquire about his or her rights and obligations. The claimant must show reasonable causes for his or her failure to enquire.	Do the facts indicate it was reasonable for such a claimant not to have enquired about his or her entitlements?
New legislation	It is normally reasonable to expect a liable person to notice advertising campaigns drawing attention to new legislation that might give rise to entitlement to benefit where there was previously none. The claimant must show reasonable causes for ignorance.	Do the facts indicate it was reasonable that the claimant did not see the publicity? In the event of a delayed claim after discovering the renewed possibility of benefit, do the facts indicate it was reasonable for the claimant not to have applied earlier?
Recently arrived from living abroad	It is reasonable that a person who arrives from abroad might not know about their potential entitlements. It is also reasonable however that such a person will inquire about entitlements soon after arriving.	Was the length of time before inquiring reasonable for the particular circumstances?
Difficulty with language	Persons are expected to seek help to find out their entitlements, such as the assistance of an interpreter. The claimant will have to show reasonable causes for their failure to do this.	Do the facts indicate that it was reasonable that the claimant did not succeed in finding anyone to help with interpretation?
Communication problems	It is normally reasonable to expect an interpreter or any other medium of communication to have operated correctly and accurately. The claimant will need to demonstrate that this did not take place.	Was the translation incorrect leading to a reasonable cause for not applying for benefit?

Box 9.1 continued

Reasons in back-dating requests	Assumption	Reasonableness tests for awarding backdate
Postal delays	It is reasonable for the claimant to suffer a degree of postal delay, either normal or unusual.	If longer than standard postal delay is claimed, is there any corroborating evidence? Was it reasonable under the circumstances that the claimant did not inquire about the progress of the claim?
Incorrect advice	It is normally reasonable for advice given by the billing authority to be correct.	Was the claimant given incorrect advice by the authority, which was acted upon and leads to delay in applying for council tax benefit? If the claimant was genuinely misadvised by a doctor, solicitor or similar professional, was it reasonable that he or she had not also take advice from the council?
Misunderstanding	It is normally reasonable to assume a claimant understands the advice given but a billing authority	Do the facts indicate there is a reasonable likelihood that the claimant genuinely misunderstand the advice given with the consequence of not applying earlier for benefit?
Carelessness, thoughtlessness or indifference	It is normally reasonable for a person to take responsibility	This in itself will certainly will not constitute good cause
Illness	A person who is ill will ask someone else to make a claim for him or her, or help complete an application, or post the application etc.	Was the claimant so ill that he or she could a claim? Has the claimant been discharged from hospital? This is accepted as a good cause of failure to claim from the day of admittance to three weeks after the day of discharge up to a maximum of three weeks.
On job seeker's allowance (JSA)	A person on JSA will reasonably apply for council tax benefit too.	Did the claimant have a genuine reason to believe that new work was imminent? Did he or she expect to be receiving pay in lieu of notice?

Calculation of benefit

Part VI of and schedules 1 to 5 to SI 1992 No 1814, as amended, provide for the calculation of council tax benefit. For the purpose of the calculation both council tax benefit and council tax liability are expressed in terms of a weekly amount. Maximum benefit is 100% of the weekly council tax liability.

Benefit is calculated with reference to a notional weekly income threshold called the applicable amount. Each claimant has their own figure for the applicable amount that is calculated by applying fixed rates for each recognised aspect of the claimant's circumstances, eg according to their age and marital status.

If the claimant's weekly income, or in the case of a couple their joint net income, is less than or equal to their applicable amount, they will be awarded maximum council tax benefit (100%)

Example 1

A married couple's council tax before benefit is £1300 for the year.
First the liability is expressed in weekly terms by dividing by 52.
This is 1300/52 = £25 per week.
Maximum possible benefit (100%) would therefore be £25 per week.
Their weekly applicable amount is calculated to be say, £150.
Their income is £140.

As their income is less than their applicable amount, they will be awarded 100% CTB. They will have no council tax to pay.

Where the claimants' net income exceeds their applicable amount, then CTB entitlement will be reduced from the maximum, by 20% of the difference. This value of 20% is know as the taper, and may be changed by central government from time to time.

Example 2

As before the couple's council tax before benefit is £1300 for the year.
First this is expressed in weekly terms by dividing by 52. This gives £25 per week.
Maximum possible benefit would therefore be £25 per week.
Their weekly applicable amount is calculated to be say, £150.
Their income however is now £200.

As their income is more than their applicable amount, their benefit will be calculated by starting with the maximum benefit and reducing it by 20% of the excess of their income over their applicable amount.

$$\text{ie } 25 - 20/100 \times 50 = £15 \text{ CTB award}$$

Their benefit award is £15 per week, leaving them with a council tax liability of £10 per week.

This can then be multiplied by 52 to find the amount of council tax left for them to pay in the year, ie £520.

Non-dependant deductions

Regulation 52 of SI 1992 No 1814 provides for non-dependant deductions. Council tax benefit will normally be reduced if a non-dependant lives in the claimants' household. A non-dependent for this purpose is generally a person who is not a partner and not a dependant child. There are some circumstances when a non-dependant deduction is not made (see Box 9.2).

Box 9.2 Where CTB is not reduced because of a non-dependant resident

The claimant or their partner is
- receiving attendance allowance
- receiving the care component of disability living allowance
- registered blind

The non-dependent is
- in detention
- severely mentally impaired
- 18 years of age or more, and not subject to child benefit payments
- student nurse, apprentice or youth
- trainee
- hospital in-patient
- staying there to be officially in care
- staying there to officially care for another
- has main residence elsewhere
- full time student
- receives income support or job seeker's allowance

Second adult rebate

Second adult rebate is a maximum 25% council tax benefit for liable persons with any income, who live with someone who is not liable for council tax (the "second adult") and who has a low income. The principle behind this is to compensate such a person for the fact that he or she is not entitled to a single person discount, even though he or she is the only occupier with more than minimal income. It may be noted that the income of the applicant is not means tested. The award is based on a means test of the second adult. To qualify for second adult rebate the second adult or adults must meet the following criteria:

- 18 years or more of age
- not liable to pay rent to anyone living in the same accommodation
- not be the husband or wife of the applicant
- not be living together as they were the husband or wife of the applicant
- not be jointly liable themselves for council tax
- have a low income (see Box 9.3 below).

A typical situation where second adult rebate applies is where a lone parent has a grown up child or children living with them who meet the above criteria.

The rates for the second adult rebate are shown in Box 9.3. It is not awarded in conjunction with other types of discount (see Chapter 13) — for example, if two jointly liable parties are already receiving a 25% discount because one of them is severely mentally imparted or is a student, then it is not possible for them to also be awarded a second adult rebate.

If a person would have an entitlement under ordinary council tax benefit and second adult rebate, then whichever of the two gives the greater award must be applied. Practitioners call this the "better buy comparison".

Box 9.3 Rates for the second adult rebate

Income of second adult(s)	Award of second adult rebate
Second adult(s) receives income support or income based jobseeker's allowance	25%
Gross income of second adult(s) is less than £137 per week	15%
Gross income of second adult(s) is between £131 and £176.99 per week	7.5%
Gross income of second adult(s) is £177 or more per week	Nil

Discretionary council tax benefit

Under the Discretionary Financial Assistance Regulations 2001 SI 2001 No 1167, council tax benefit may be awarded by the billing authority at its discretion where an applicant is experiencing exceptional circumstances, and is not otherwise entitled to council tax benefit. The award can be up to 100% of the applicant's council tax liability. In practice, this kind of benefit is seldom awarded and difficult for an applicant to secure. This is because local authorities receive no subsidy for payments of discretionary benefit. Also, the sum total of an authority's discretionary council tax benefit awards must not exceed 0.1% of the total council tax benefit paid within that year. Applicant's circumstances must therefore be very exceptional and include hardship. One example sometimes quoted by practitioners to give an idea what is exceptional is a person who has been burgled, had their savings stolen and is enduring extreme hardship as a result.

This kind of benefit can be a one off award, or awarded on an ongoing basis while the exceptional circumstances remain. There should be no predefined conditions for the purpose of assessing a claim. Each case is to be considered on its individual circumstances. Some authorities that make such awards may be very careful not to commit the full permissible budget early in the year, as this would leave no funds in the eventuality of an even more worthy case emerging later in the year.

Compensation refund scheme

Where a person has received certain benefits, including council tax benefit, and subsequently becomes eligible for compensation, the compensator must have regard to the compensation refund scheme (CRS) which is run by the DWP. Briefly, a notification and payment procedure requires, step by step:

- the compensator to notify the DWP of the pending payment of compensation
- the DWP informs the compensator of the amount and kinds of benefit which must be repaid directly to the DWP
- the compensator pays the amount to the DWP and pays compensation to the claimant less the amount paid to the DWP.

Benefit "appeals"

A claimant who is dissatisfied with the council tax benefit decision and wishes to dispute it may:

- question the billing authority about its decision and ask for the claim to be rechecked
- appeal to the benefit appeals service under section 68 and schedule 7 to the Child Support, Pensions and Social Security Act 2000
- appeal to the social security commissioners on a point of law.

In Chapter 17 on disputes these approaches are considered in some detail.

Council tax benefit in Scotland

Council tax is commonly applied throughout Great Britain so the levying authorities operate the scheme in Scotland.

If the Council Tax Abolition and Service Tax Introduction (Scotland) Bill had been adopted by the Scottish Parliament the need for council tax benefit would have been removed from the date when service tax came into force. It may be noted that the proposal intended that all outstanding liabilities for council tax would have been removed from 1 April 2006. If this were to have happened, there should have been no hang over of council tax benefit.

Counter-fraud Practice

Aim

To review the policies and practice of counter fraud work, particularly by the billing authorities

Objectives

- **to describe the "joined up" policies and practice of counter-fraud work in society**
- **to give a schematic organisation and structure for this work in a billing authority office, including roles and activities**
- **to describe the offences and the investigatory policies and practices in countering them, with reference to the main statutory provisions**

Introduction

Every local authority is required to have an anti-fraud policy which covers all departments. In this chapter the subject is only considered from the council tax and council tax benefit perspective.

Organisation and structure

A billing authority is required to have a written policy on anti-fraud and anti-corruption. Performance indicators show the nature of the requirements for administration in this function.

The organisation and structure for anti-fraud and anti-corruption will, of course, vary from authority to authority but generally the requirements will be covered by the following:

- recruitment of staff
- induction of staff to cover policy and practice on anti-fraud and the like
- general and specific training to detect internal and external problems
- guidance on procedures for reporting problems
- specific roles and activities allocated to personnel with appropriate training, development and accreditation where appropriate

- procedures for handling internal and external allegations of an offence having been committed
- liaison and partnership arrangements with third party organisations to pursue alleged offenders with the view to prosecution.

Policy for counter-fraud

Most billing authorities should have a written counter- or anti-fraud policy as recommended by government. It should cover both internal and external fraud, covering such matters as:

- the conduct of staff in being vigilant
- procedures for staff to handle and report incidents of suspected fraud
- the adoption of the Public Interest Disclosure Act 1998 to protect staff and others who report findings
- measures for the prevention of fraud, eg by thorough assessment of claimants
- the deterrence of fraud, eg by publicity of successful prosecutions
- the detection of fraud, eg by use of the joined-up counter-fraud facilities available from outside agencies
- the promotion of publicity on counter-fraud policy, eg placing the policy on the authority's website.

All members of staff and elected members might be expected to receive a copy of the policy and sign for it. Regular updates should be provided to them when a change is introduced — again recipients should sign for them.

Roles and activities

The roles and activities for the prevention of fraud and corruption in a billing authority and other organisations are listed in Box 10.1. The box illustrates the increasingly joined-up nature of the national counter-fraud operations.

Joined-up working arrangements

A key element of anti-fraud work is the way in which the billing authority and other bodies act in concert on a range of matters (see Box 10.1).

Mail interception

The Royal Mail assists in anti-fraud practice by providing a service to billing authorities. It intercepts benefits' mail which has been redirected in one of the two ways available to claimants and others namely:

- the official Royal Mail redirection service
- informal redirection by re-addressing the mail from the first address (which has been arranged by the addressee or another).

The mail is sent by Royal Mail to the billing authority concerned.

Box 10.1 Roles and activities in the investigation and prevention of fraud

Department for Works and Pensions	• departmental strategy — *Reducing fraud in the benefit system*
• Fraud Investigation Team	• team with powers of investigation and penalty
• Housing Benefit Matching Service	• matches information on housing benefits and council tax benefits with other national benefits records
	• assists detection of fraud by identifying mismatched records
• SOLP	• a Solicitors Practice scheme — free to local authorities on service level agreements
• National Fraud Hotline	• a telephone hotline on fraud
	• calls may be referred to the local billing authority
• SAFE scheme	• Security Against Fraud and Errors contributes towards billing authorities' various costs of counter-fraud work, eg in housing benefit
Home Office	• possible provision of travel data to detect fraud
Local Authority Investigation Officers Group	• National body of representatives of billing authorities
Audit Commission	
• National Fraud Initiative	• a matching service covering many databases
	• enables searches by national insurance number(NINO), name, date of birth and other items
Royal Mail Corporate Security Criminal Intelligence	• intercepts redirected benefits mail
	• assists in the identification of person arranging redirection
Billing authority	
• audit officer	• officers with various titles who undertake counter-fraud operations
• fraud manager	• surveillance, data matching, interviews (some under caution)
• senior investigation officer	• prosecutions activities and so on
• disclosure officer	
• recovery officer	
• interventions visiting officer	

Also, where the billing authority wishes to trace the claimant who has moved address the Royal Mail will seek to furnish details of the address used for the redirected mail.

Induction, training and development

At induction all staff (and members) of local authorities should receive a presentation about their authority's anti-fraud and anti-corruption policy. A copy of the policy would normally be signed for when it is given to them. From time to time reminders and or updates will be issued during the year.

Officers with responsibility and duties in the field will undertake special occupational development, eg prospective investigation officers will need training in working to the requirements of the Police and Criminal Evidence Act 1984. The developmental activities are almost certain to lead to some of them gaining specialist qualifications, such as:

- professionalism in security (PINS)
- qualification as an accredited counter fraud manager (senior investigation officer)
- qualification as an accredited counter fraud officer (benefit investigation officer).

Senior officers in this field of local government may represent their authority in the Local Authority Investigation Officers Group (LAIOG).

Verification framework (VP)

The verification framework is an investigatory methodology developed nationally (which is voluntarily adopted by billing authorities) to establish that a claim is true. The funding for setting up a VF is due to end in April 2006 and applications for it will be received up to 3 March 2006. Ongoing funding remains. The key elements include the following:

- the claimant proves his or her identity and provides their national insurance number
- the claimant supports a claim by supplying original genuine documentary evidence
- the investigator will verify that none of the documents have been tampered with, eg by using an ultra-violet scanner
- any changed circumstances will be checked
- the information supporting any previous claims will be checked against that supplied with the current claim.

Direct payments

Until recently successful claimants received an order book which was used at the Post Office to claim council tax benefit and other benefits. As a counter-fraud measure, payments are now made directly to an account set up by the claimant, eg at the Post Office.

Investigations

Code of conduct in fraud investigations

As well as a policy for counter-fraud, billing authorities are encouraged to have a code of conduct in fraud investigators. It should include:

- the need to comply with all legislative requirements and other legal matters (see below)
- behaviour towards suspects and potential witnesses while conducting investigations
- the appropriate ways to handle elderly individuals, persons with mental health problems and minors
- for visits, the appropriate notification, timing and conduct, eg visiting the above and those who live alone, for instance, a woman.

Box 10.2 Investigatory powers and associated legislation

Criminal Procedure and Investigation Act 1996	• deals with procedural matters
Data Protection Act 1998	• concerned with the improper use of information about an individual
Human Rights Act 1998 • concerns a variety of rights • three particularly relevant	• right to a fair trial or hearing • right to privacy • right to not be discriminated against
Police and Criminal Evidence Act 1984 (PACE)	• concerns protected disclosure of fraud at work
Public Interest Disclosure Act 1998	• provides for protection of "whistleblowers"
Regulation of Investigatory Powers Act 2000	• procedures for surveillance • must be authorised by relevant senior officer
Social Security Fraud Act 2001 • billing authorities and others have the right to obtain information	• provides for the exchange and disclosure of information by electronic and other means • relates to many kinds of businesses, financial institutions and other bodies who may have dealings with those suspected of benefits fraud

Complaints procedures

The above policy should include the promotion of the authority's complaints procedure and the availability of other avenues for complaints (see Chapters 17 and 18).

Legislation

Much of the legislation concerning investigatory practice in anti-fraud and associated matters concerning council tax and council tax benefit is reviewed in Box 10.2. The powers should enable adequate investigations by the appropriate authorities and at the same time have due regard to the individual's rights.

Tracing agents

Tracing agents may be used by billing authorities to search for and provide forwarding addresses for liable parties who have absconded.

Surveillance policy and practice

Generally, surveillance of a suspect is likely to be an approach of last resort and in keeping with the nature of the supposed offence. A billing authority will have a policy and adopted method of surveillance. Essential features are likely to include:

- it must comply with statutory requirements, ie the Regulation of Investigatory Powers Act 2000 and code of conduct
- surveillance investigators must be suitable for the work
- every surveillance must be appropriately authorised by an accredited senior officer
- the actual work should not interfere with the rights of third parties and end once the evidence has been obtained or when it is considered that it will not be forthcoming
- notes, photographs and other collected item should be kept for a period, eg at least three years.

In *R v Turnbull* [1978] 3 All ER 549 the conduct of surveillance practice was itself under scrutiny. The case provides an insight into the kind of evidence those carrying out surveillance might be expected to provide in court, including:

- the duration and time of day
- the distance, line of sight, obstructions, eg traffic, and weather for visibility
- prior sighting of the suspects and others under surveillance
- duration of time between the surveillance and recording of the happenings
- explanation of errors or important discrepancy in the report
- the rationale of the findings of the surveillance.

Interviews

The interview under caution (IUC) is carried out in accordance with both the Police and Criminal Evidence Act 1984 and code of practice of practice number 7 under the Act. The words of the caution are:

> You do not have to say anything. But it may harm your defence if you do not mention when questioned something which you later rely on in court. Anything you do say may be given in evidence.

Suspect's rights

The suspect has many rights which must be observed by the billing authority's investigators, the police or others with powers to investigate. The rights arise under the following sources:

- common law, eg the law of natural justice (to the extent that it has not been superseded by statute)
- statutes, eg the Human Rights Act 1998 (see Box 10.3)
- policies incorporated in codes of conduct and other material.

Human rights

The Human Rights Act 1998 requires that an individual's rights are not breached by the billing authority and other public parties to an investigation. The rights which may be particularly relevant to the issue are:

- the right to liberty and security
- the right to privacy and family life
- the right to a fair trial.

Data protection

The eight principles for the protection of an individual's right to privacy embodied in the Data Protection Act 1998 are given in Chapter 5 (see p52). In carrying out investigations the investigation officers must have regard to the principles and codes of conduct drawn up under them.

Hierarchy of sanctions and penalties

There are essentially three levels of sanction in anti-fraud work, that is:

- the formal caution
- the administrative penalty
- prosecution

Appeals against penalties are made to the valuation tribunal.

Offences

Three main bodies of legislation have provisions covering offences committed in the context of council tax. Box 10.3 sets out the sources and nature of the offences.

Defences

Possible defences against a charge that something was stolen include:

- honest belief that in law it could be taken
- took reasonable steps to inform the owner but thought the owner could not be found
- borrowed it with the intention of giving it back.

Formal caution and administrative penalty

The ultimate penalty is perhaps prosecution and conviction by a magistrate under a case taken by the billing authority under regulation 65 of the Council Tax Benefit (General) Regulations 1992 SI 1992 No 1814. However, the billing authority has other penalties which it may use, namely:

- "formal caution" under the Social Security Administration Act 1992, as amended, which is entered on a national benefits database — should a further offence come to light it may be considered
- "administrative penalty" under the same Act, as amended, which is a fine of 30% of the amount of council tax benefit which was overpaid.

Box 10.3 Some offences in connection with council tax and council tax benefit

Social Security Administration Act 1992 s 112(1)(a)
- making a false statement or representation
  - 3 months imprisonment for each offence up to a maximum of 6 months
  - fine up to £5000

Social Security Fraud Act 2001 s 16 and
Social Security Administration Act 1992 s 112
- failure to give a prompt notification of a change effecting any benefit entitlement
  - 3 months imprisonment for each offence up to a maximum of 6 months
  - fine up to £5000

Theft Act 1968 s 15 (England and Wales)
- dishonestly obtaining property by deception
  - If heard summarily, 3 months imprisonment for each offence up to a maximum of 12 months
  - fine up to £5000
  - If heard on indictment, up to 10 years imprisonment and unlimited fine

Social Security Administration Act 1992 s 112(1)(b)
- allowing information to be provided knowing it is false
  - 3 months imprisonment for each offence up to a maximum of 6 months
  - fine up to £5000

Social Security Fraud) Act 1997 and
Social Security Act 1992 s 111A
- dishonestly making false statement or representations
  - If heard summarily, 3 months imprisonment for each offence up to a maximum of 12 months
  - fine up to £5000
  - If heard on indictment, up to 7 years imprisonment and unlimited fine

Social Security Fraud Act 1997 and
Social Security Administration Act 1992 s 111A
- dishonestly allowing information to be provided knowing it is false
  - If heard summarily, 3 months imprisonment for each offence up to a maximum of 12 months
  - fine up to £5000
  - If heard on indictment, up to 7 years imprisonment and unlimited fine

Social Security Fraud Act 2001 and
Social Security Administration Act 1992 s 112(1)(b)
- failure to give a prompt notification of a change affecting any entitlement of another person
  - 3 months imprisonment for each offence up to a maximum of 6 months
  - fine up to £5000

Social Security Fraud Act 2001 and
Social Security Administration Act 1992
s 112(1)(c)–(1)(e)
- failure to notify a change, or causing or allowing someone to fail to notify a change
  - 3 months imprisonment for each offence up to a maximum of 6 months
  - fine up to £5000

Theft Act 1968 s 17
- false accounting
  - If heard summarily, 6 months imprisonment for each offence up to a maximum of 12 months.
  - fine up to £5000
  - If heard on indictment, up to 7 years imprisonment and unlimited fine

---

Box 10.3 continued

Theft Act 1968 s 15A and Theft (Amendment) Act 1996 • dishonestly obtaining a money transfer	• If heard summarily, 6 months imprisonment for each offence up to a maximum of 12 months • fine up to £5000 • If heard on indictment, up to 10 years imprisonment and unlimited fine
Theft Act 1968 s 24A and Theft (Amendment) Act 1996 • dishonestly retaining a wrongful credit	• If heard summarily, 6 months imprisonment for each offence up to a maximum of 12 months • fine up to £5000 • If heard on indictment, up to 10 years imprisonment and unlimited fine

---

Again under the same Act, the billing authority may refer a case to the police for action, eg a "police caution" which is entered on the National Police Computer.

"Two strikes" sanction

From 1 April 2002, a person who is convicted on at least two occasions within three years of each other for different benefit offences may have the "two strike" provisions of the Social Security (Loss of Benefit) Regulations 2001 SI 2001 No 4022 under the Social Security Fraud Act 2001. In effect the individual has a sanction on the payment of benefit during a period of disqualification for benefits.

Voluntary disclosure

The Department for Works and Pensions encourages those in receipt of council tax benefit who commit fraud to fully disclose it voluntarily. Unless the fraud is serious, prosecutions are not normal after such disclosures. This may apply to others who commit fraud even though they are not actually in receipt of benefit. However, those who commit fraud and disclose it "voluntarily" during the course of an official investigation and certain other circumstances will not be treated as having made a voluntary disclosure.

Billing authority rewards

Successful counter-fraud measures by a billing authority will normally result in recognition and rewards for the authority under the SAFE scheme (see Box 10.1).

A fixed amount is paid for every sanction that the billing authority makes, ie for every formal caution, administrative penalty and issuance of a summons. Furthermore, a sum is paid for any prosecution which is successful.

Best Value and Performance

11

Aim

To explain how the performance of billing authorities is assessed in the comprehensive performance assessment regime and other means

Objectives

- to identify the organisations which assess a billing authority's functions and those of other bodies and other
- to define the terminology of best value and performance
- to outline the approach to improvement of performance
- to examine the roles and activities of the principal participants
- to indicate the future

Introduction

The White Papers *Modern local government: in touch with the people* and *Strong local leadership: quality local services* initiated the government's policy on the best value approach to the assessment and review of the ways in which local government deliver their services to the community.

It could be said that local government members are continuously assessed on their performance by the electorate who vote them in and out of office. However, this chapter deals with the approach to the performance and assessment of revenue services concerned with council tax and council tax benefits. In general, the aim of the work is to increase efficiencies in public services and thereby give value for council taxpayers' money. (The authors are conscious that they are aiming to explain a moving target (see *Future* on p129.)

Roles and activities

The area covers such bodies as:

- the Audit Commission

- the Benefit Fraud Inspectorate
- Best Value Inspectorate
- comprehensive performance assessment inspectors.

Other aspects of "performance" result from the deliberations and decisions of several other bodies including the following:

- the several ombudsmen involved in maladministration
- the tribunals and courts.

The roles and activities of these bodies are dealt with in Chapters 17 and 18.

Many of the roles and activities in the assessment of performance are conducted by national inspection agencies — some of the 11 are shown in Box 11.1. (It may be noted that they are to become four.)

Box 11.1 Roles and activities in performance and targeting at national level

Best Value Inspectorate Forum
(comprises the heads of the Inspectorates)

- meet two or more times a year
- coordinate their work
- seek to jointly inspect etc
- promote best practice
- advises government on policy
- reviews policy and practice

HM Inspectorate of Constabulary
HM Fire Service Inspectorate
The Office of Standards in Education
The Social Services Inspectorate
The Best Value Inspectorate

- formulate best practice standards
- carry out inspections
- carry out special investigations
- prepare reports
  (taking in the Housing Inspectorate)
- seek action by local bodies to improve any deficiencies

Benefit Fraud Inspectorate

- (as for other inspectorates)
- publishes a guide to good practice on council tax benefit and housing benefit

---

At a local level the managerial response to the statutory arrangements for performance assessment is often embedded in the line managers' roles and activities, perhaps assisted by staff with special responsibilities for managing the processes throughout the local authority.

Staff will be aware that the formalised approach to best value and performance is embedded in the work of the Audit Commission and individual inspectorate but there is another dimension — the courts, tribunals and ombudsmen. As soon as a decision by one of these bodies is delivered a review should pick up any adverse aspects of best value and performance and incorporate it, if appropriate into the formal procedures.

Operational performance management
Audit Commission

The Audit Commission's role and activities are established under the Local Government Act 1999 and for Wales, the Wales Audit Office by the Public Audit (Wales) Act 2004. Its work covers the audit and performance of public bodies. Whereas each of the specialised bodies in Box 11.1 tends to be concerned with one or a limited number of functions, the Audit Commission's role is assessment and reporting on performance in an overarching sense. It can even include assessments of other official bodies in arriving at overall assessments for a particular public organisation. For example, in *R (Ealing LBC)* v *Audit Commission for England and Wales* (2005) Times May 26, the court held that the Audit Commission could, in effect, delegate its function of decision-making. The court upheld the Audit Commission's right to include in its comprehensive performance assessment of Ealing London Borough Council the very bad score given to the council by the Social Care Inspectorate. This happened to result in the council comprehensive performance assessment (CPA) (see below) being downgraded from "good" to "weak". (The Audit Commission publishes or updates working documents as a basis for its approach, eg *Comprehensive Performance Framework 2005* (see p127).)

Best value

Best value was established by Part I the Local Government Act 1999 (1999 Act) and came into force on 1 April 2000 — partly to ameliorate competitive tendering. The 1999 Act imposes a duty:

- to plan for improvements in a prescribed way
- to demonstrate performance with reference to some key performance indicators and their associated standards
- to demonstrate continuous improvement with reference to economy, efficiency and effectiveness of functions.

The last requirement is particularly important.

Under sections 6 and 7 of the 1999 Act each authority must maintain audited best value performance plans (BVPP) (see below) for each of its services. It must also publish performance reports. Any of these documents may inspected by the Best Value Inspectorate.

In compiling a BVPP an authority is expected to demonstrate what is sometimes referred to as the four Cs of best value, namely:

- *Challenge*, such as asking if a particular function needs to be carried out in a particular way, or at all
- *Compare*, to other authorities, and similar service providers both in the public and private sector
- *Consult* with stakeholders, such as the public and councillors regarding their view of services and aspirations for the future of services
- *Compete*, where necessary and appropriate by offering services for competitive tender.

The last, *Compete*, is seldom invoked in revenue services on a large scale, although it is retained as an ultimate sanction for services that continue to perform poorly, or for whom less dramatic intervention can turn performance around. However, having said that, an outsourcing exercise to deliver a function — with internal staff competing against a private sector contractor — would come within the ambit of *Compete*. (The four Cs are particularly important for the review of best value.)

Performance indicators

Section 4 of the 1999 Act provides for performance indicators and performance standards. At the moment there are 97 best value performance indicators (excluding those concerning the fire service). Of these, 10 relate directly to:

- council tax
- housing benefit and council tax benefit
- fraud concerning housing benefit and council tax benefit.

The Best Value Inspectorate has an inspection plan which examines every local authority's performance statistics (Stats). Stats for 31 March 2004 are shown in Box 11.2.

Box 11.2 Performance indicators for revenue purposes

Indicator No	Description	National Average 2003–2004
BV9	• council tax collected (as a %)	
BV76a	Housing Benefit Security HBS) • claimants visited per year per thousand	215.7
BV76b	HBS • investigators per 1000 caseload	1.0
BV76c	HBS • number of investigations per 1000 caseload	43.7
BV76d	Prosecutions and sanctions • per 1000 claimants	3.8
BV78a	Average time (AV) • for processing new claims	45.5 days
BV78b	AV • for processing change of circs	13.7 days
BV78c	Renewal claims • % processed on time	66.6%
BV79a	Accuracy • % of cases calculated correctly	96.9%
BV79b	Recoveries • % of overpaid housing benefit and council tax benefit recovered	47.7%

Best value and performance

The White Paper *Modern Local Government; in touch with the people* of 1998 provided the basis of policy for best value. Best value represents a system of plans, priority areas, targets and performance indicators. A monitoring of best value practice is carried out by the Audit Commission. A number of topics are selected from time to time and the best achieving individual authorities are picked for "Beacon Status".

Best value performance plan

A best value performance plan (BVPP) is prepared for stakeholders by a local authority or other body for the following purposes:

- submission to the government as required by law
- as a source of information available to the public
- as a source of information for suppliers and others
- as a foundation for members and staff to manage performance.

It may comprise one document but could comprise two or more. It is likely to include:

- actual or summary plans for functions or services
- within the plans, performance indicators, targets and programmes
- financial accounts
- references to national and other standards underlying the individual plans
- details of the performance of peer bodies — essentially as a basis for comparison.

The plan is intended to demonstrate to the government, the local authority's commitment and progress to good value and at the same time its accountability to the public. The BVPP should, therefore, be promoted and written in a style (with a summary) which engages the stakeholders, particularly the members of the public. It is audited each year.

Use of performance indicators

The government uses 97 best value performance indicators to cover the work of local authorities. Of course, most of these do not relate directly to council tax and council tax benefit administration (see Box 11.2 on p124). However, the regime for performance assessment requires a hierarchical approach to action in this field. For a given service, public bodies are required to establish "priority areas" in the prescribed "targets". Essentially, the targets are established for a particular year as the basis of the "Best Value Performance Plan".

The performance indicators may be used in various ways, including:

- as a comparator with the previous years — thus showing any trend
- as a comparison with the performance indicators of other similar bodies — a peer benchmarking approach
- as a basis for review and analysis of operations.

Council tax performance indicators

The performance indicators for council tax which are used to evaluate local authorities are:

- the amount of council tax collected as a percentage of the amount which could be collected for the year
- the average annual cost for a dwelling of collecting the billing authorities' council tax.

Priority areas

Each priority area is measured on the following:

- cost/efficiency
- service delivery outcomes
- quality
- corporate health.

A priority area may be identified from a number of sources, such as:

- the need to meet the requirements of government standards which have just been published
- a CPA presentation identifying an area of work with a below average performance
- a Benefit Fraud Inspectorate report on the administration of council tax benefits.

Beacon status

Each year various individual services of particular local authorities are awarded the accolade of "beacon status". As a result each local authority's service is promoted nationally, regionally or locally as an example of best practice. Officers of the authority may be invited to publish details of their approach and to participate in training conferences and other events.

Relationship between best value and comprehensive performance assessment

Best value is an obligatory method of performance planning and measuring. Comprehensive performance assessment (CPA) was introduced in 1992. It is methodology which measures performance in a way that is different to best value. Best value inspectors usually review the best value plans of one or two services within an authority. CPA is an assessment by the Audit Commission of the authority as a whole.

The CPA approach does not prescribe a method for performance planning or periodic measurement. However, it does impose the following;

- a requirement to follow up on the weaknesses identified in the process
- an incorporation of appropriate improvements into the BVPP.

CPA is in line with the general principles of best value and the outcomes of a CPA inform the best value process. For example, an authority that has good or excellent scores in their CPA will have less best value inspections. Those with poorer CPA reports are likely have an increase in best value inspections.

Both best value inspections and CPAs can lead to intervention where there is poor performance.

Comprehensive performance assessment

The Audit Commission conducts CPAs of local councils and other public bodies, with additional emphasis on the customers' point of view and community matters, eg council tax and council tax benefits performance may be as important as any other service. There are two aspects of the approach, namely:

- corporate assessment — for instance, ambition for new developments, focus on improvement and leadership in partnership
- services assessment — for instance, value for money, efficiency and cost effectiveness of key services, including education, environment, housing, libraries, leisure services, social services and the use of resources.

The work is carried out in line with its principles of performance measurement. Briefly, they may be listed as:

- "clarity of purpose" — for the users of the information
- "focus" — essentially on the priorities of the body
- "alignment" — of targets, the use of performance indicators and the objective-setting of the body
- "balance" — in terms of use of indicators to performance and the cost-effectiveness of the information collected
- "regular refinement" — performance indicators should reflect the current requirements
- "robust performance indicators" — day to day usage of data is important as well as internal and external monitoring.

A full statement and exploration of the principles is given in *On target: the practice of performance indicators*, a management paper by the Audit Commission.

The process classifies the performance under quality of service, eg "good", and gives a separate classification for likelihood of improvement, eg "likely" (see Box 11.3, but these are likely to change).

Box 11.4 shows schematically an approach to a CPA which extends the detail to 12 steps.

Internal targets

Each service within a local authority, including revenue services, will have an annual business plan. This will include locally developed aspirations for performance, quality and service development, as well as targets derived from best value and CPA.

Senior management is always responsible for most aspects of performance, though some authorities have a corporate official who supports or advises on performance, sometimes known as a "performance manager".

Box 11.3 Comprehensive performance review — performance classification
(subject to consultation these are likely to change (see p129))

Quality of Service		Likelihood of Improvement
Excellent	(3 stars)	Yes
Good	(2 stars)	Likely
Fair	(1 star)	Unlikely
Poor	(0 star)	No

Box 11.4 Schematic comprehensive performance assessment of a service

Step	Local authority	Other	Inspectors
1	• Receives notification of forthcoming CPA • Prepares self assessment		
2		• Peer challenge by consultancy group with independent members	
3	• Prepares document inspection pack • Sends it to the inspection team		
4			• Receives documentation pack • Team familiarises itself with the material
5a	• Greets team • Accommodates them with office and telephones etc		
5b			• Visits council • Undertakes inspection • Does reality checks
6			• Prepares draft interim report
7		• Team reviews draft with other Audit Commission inspectors	
8a			• Finalises the interim report • Sends copy to LA preview team for interim challenge
8b	• LA pre-view officers comment back on interim challenge		
9			• Prepares final report
10	• Receives the presentation		• Presents the final report to the council and senior
11	• Inform staff of the report • Reviews the findings • Develops initiatives to address weaknesses • Seeks to improve performance overall		
12	• Implements and monitors initiatives		

Gershon efficiencies

A renewed emphasis on cost efficiency is affecting local government including revenue departments following an independent review of public sector efficiency by Sir Peter Gershon published in July 2004. The recommendations and expectations of this review were largely adopted by government. These require a saving in public sector expenditure of £20 billion between 2004 and 2008. Savings of £6 billion savings are to be found by an intended increased efficiency in local government. This represents around a 2.5% reduction in costs. As a result measurements, such as "cost of council tax collection" and "cost per benefit claim", may return in the form of major performance indicators.

Benefit Fraud Inspectorate

Section 14 of the 1999 Act makes special mention of inspections of the practices concerning housing benefit and council tax benefit. It also refers to benefit fraud. This is done by substituting new sub-sections into section 139A of the Social Security Administration Act 1992. It may be noted that from time to time a local authority may be visited by the Benefit Fraud Inspectorate. Details of the approach by the BFI are given in Chapter 13.

Guide to good practice

The Benefit Fraud Inspectorate has published the "Guide to Good Practice" for the local authorities concerned with council tax benefits. (It also covers housing benefits.)

Future

As indicated in the *Introduction* above the framework for improvements in local government is a moving target. Thus, Audit Commission publications on comprehensive performance assessment for the years to 2009 are out for consultation towards the end of 2005. They include:

- *The Framework for Comprehensive Performance Assessment of District Councils for 2006*
- *The Harder Test — Single Tier and County Councils' Framework for 2005.*

However, this operational future may be considered in the wider reform of local government and the inspection of its services. Pressures on local authorities are developing rapidly, including:

- the changing nature of the government's policy on inspection
- the forthcoming recommendations of the Lyons Inquiry
- the impact of Gershon on cost efficiency
- the overhaul of inspection serevices, including the intended focus of the inspectorate for local services.

As mentioned in the Budget Statement 2005 by the Chancellor of the Exchequer, an inspectorate for local services will be one of the four new inspectorates created from the existing 11. It seems certain to be a joining of the Audit Commission with the Benefit Fraud Inspectorate — as described in the ODPM's consultation paper of 28 November 2005, ie *Inspection reform: the future of local services inspection.*

Part 5

Council Tax Payers

Council Tax Payers

12

Aim

To identify those who are liable for council tax

Objectives

- **to identify and describe occupiers who pay council tax**
- **to identify owners who are liable to pay council tax**

Introduction

This chapter explores who pays council tax, ie occupiers and in some instances owners (as landlords or others). Other chapters deal with the following associated topics:

- the ways and means of making payments (see Chapter 8)
- exemptions, deductions, discounts and other reliefs (see Chapter 13)
- the resolution of disputes is covered in Chapters 17 and 18.

There are about 22 million persons liable to pay council tax in respect of dwellings (compared with about 41 million liable to pay the former community charge or "poll tax" of the early 1990s). The billing authority determines the liable party for each dwelling in accordance with legislation that aims to deal with every scenario. Normally, the adult occupying holder of the dwelling is liable to pay the tax unless specifically exempt. All the liabilities mentioned below may be subject to one or more of a number of discounts, exemptions or benefits that reduce the amount to be paid. These are discussed in detail in Chapters 9 and 13. The year's council tax is apportioned on a daily basis, for example where a person becomes or ceases to be a liable person.

Occupiers, landlords and others

Although council tax is essentially a tax on the occupier of a dwelling, a more precise examination divides taxpayers into occupiers or owners not in occupation. The Act gives a hierarchy (see Box 12.1).

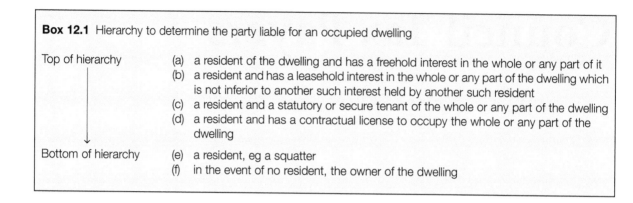

Box 12.1 Hierarchy to determine the party liable for an occupied dwelling

Top of hierarchy

(a) a resident of the dwelling and has a freehold interest in the whole or any part of it
(b) a resident and has a leasehold interest in the whole or any part of the dwelling which is not inferior to another such interest held by another such resident
(c) a resident and a statutory or secure tenant of the whole or any part of the dwelling
(d) a resident and has a contractual license to occupy the whole or any part of the dwelling

Bottom of hierarchy

(e) a resident, eg a squatter
(f) in the event of no resident, the owner of the dwelling

This list is given in section 6(2) of the 1992 Act. They are persons liable to pay council tax. Of course, there are those who are exempt or enjoy relief from the tax (see Chapter 13). Also, there are some dwellings which are exempt (see Chapter 13). In *Hammersmith & Fulham Billing Authority v Butler* [2001] RVR 197 the date of notification to the taxpayer was held not to be the date for liability to begin but liability from the date when the dwelling was so designated as being within the ambit of council tax.

Owner occupiers

Apart from a few prescribed cases where non-resident owners are liable for council tax on occupied properties (see Box 12.3), owner-occupiers are usually liable for council tax on their sole or main residence. Often there may be more than one adult in a dwelling with an interest in the property. In this case the liable party or parties are determined with reference to a hierarchy of persons given in section 6(2) of the 1992 Act.

The occupier who is the most senior in the hierarchy will be the one who is liable for council tax. If the most senior in the hierarchy shares that level with others, they will be jointly and severally liable unless they are severely mentally impaired or a student.

Joint and several liability

For any joint liability, section 9 of the Local Government Finance Act 1992 provides that liability for council tax is always "joint and several". This means that for spouses, both parties continue to be responsible until the entire liability is paid. A party can not relinquish themselves of liability by paying "their half", for example. The local authority is likely to enforce against whichever party from whom they can collect most quickly and economically.

Spouses and civil partners

Spouses are joint and severally liable for council tax under section 9 of the 1992 Act. Section 140 of the Civil Partnership Act 2004 amends section 9 of the 1992 Act and thus provides from 5 December 2005 for those in a formal civil partnership to be treated in the same way as married partners for the purpose of determining joint and several liability. The exceptions are a spouse or civil partner who is severely mentally impaired or a full time student. These are not held jointly or severally liable.

The 1992 Act provides that persons living as husband or wife (but not in a formal civil partnership) will be jointly and severally liable, even if they are not at the same level in the hierarchy.

Exclusion of Crown exemption

Traditionally, the Crown has been exempt from taxes but in recent years changes have been introduced. As far as council tax is concerned statutes particularly section 19 of the 1992 Act, excludes the exemption from certain kinds of property and from persons associated with the property. The property concerned is that maintained for Crown purposes by specified authorities including the following:

- a billing authority
- a county council
- certain police authorities, including a combined police authority under the Local Government Act 1988.

Sole or main residence

Box 12.1 refers to a person being 'resident'. However, it is not always clear if a person is in fact resident, for example, a person with two homes or a person who spends time on board a ship. For the purposes of determining the liable party for council tax, a person with more than one abode is considered to be "resident" at their main residence. Box 12.2 highlights some of the case law on "sole or main residence" for council tax purposes.

Subject to any exemptions and reliefs, if both homes are in Great Britain the main residence will attract full council tax and the second home will attract a discount of from 10% to 50% as though it were unoccupied (provided it is nobody else's sole or main residence) (see below: *Owners of unoccupied or vacant property*). The judiciary has grappled with the definition of main residence and there remains sufficient ambiguity for there to be many borderline cases. Generally the local authority will consider several factors arising from case law, such as:

- where the majority of a person's possessions are kept
- where their family lives
- where they have a bank account
- where they are registered with a doctor
- where their correspondence is delivered
- where they spend the majority of their time.

Case law also directs that no single one of these factors should be considered as overriding. In *Bennett v Copeland Borough Council* [2004] EWCA 672 the Court of Appeal decided that an individual who owned the freehold of a cottage but had never lived in the property, could not be said to have it as his sole or main residence in accord with section 6(5) of the 1992 Act.

Box 12.2 Selection of cases on the concept of sole or main residence

Individuals at sea	*Bradford Metropolitan Borough Council* v *Anderton* [1991] RA 45 Merchant seamen spent about 70 days a year at residence where his wife lived — was determined as his main residence: he had his home there
Wife and children	*Cox* v *London (SW) V&CC Tribunal Registration Officer* [1994] RVR 71 Held: reasonable in some instances to take the place where a man's wife and child resided as being his main residence
Security of tenure	*Ward* v *Kingston-upon-Hull* [1993] RA 71 Held: although away abroad for over 40 weeks a year, main residence was where security of tenure was better, ie in Hull
Security of tenure Reasonable onlooker No one factor	*R (Re Williams)* v *Horsham District Council* [2004] EWCA Civ 39 Held: tied accommodation from employer may be main residence if it would be thought reasonable so to a reasonable onlooker and that no one factor should be considered to exclude other factors

---

Receiver with enduring power of attorney

When a council tax payer (or recipient of council tax benefit) is incapacitated and cannot manage their affairs, problems may arise with the payment of council tax and other matters. In these circumstances, the Enduring Powers of Attorney Act 1985 provides that the Court of Protection may grant, for instance, a relative or close friend enduring power of attorney (EPA); they are known as the "receiver". The receiver may deal with the individual's affairs as if in their shoes, subject to insuring in a bond, keeping an account and arranging for an official appointee, the Visitor, to see that the individual is being looked after appropriately. The Visitor is appointed by the Lord Chancellor.

Personal representatives

Under regulation 58 of the Council Tax (Administration and Enforcement) Regulations 1992 SI 1992 No 613 a deceased's executor or administrator is responsible for the following;

- any council tax liability
- an unpaid penalty
- any sum due back (being in excess of lifetime liability)
- supply information (on sufferance of penalty under regulation 58(1)(c)).

The deceased estate is liable for the first two and the last is recoverable. The personal representative is only liable in that capacity.

Where proceedings were in hand at the date of death, the personal representative may continue them or withdraw from them. If necessary he or she may commence proceedings.

Tenants

It can be seen from Box 12.1 (p134) that an occupying tenant is liable to pay council tax unless there is a more senior leaseholder or owner in occupation of the same dwelling or unless the house has been designated a "house in multiple occupation" (see below). The terms of the tenancy may, however, provide that the non-resident landlord will pay the council tax. This is merely a contractual obligation. So, if the non-resident landlord does not pay the council tax as agreed with the tenant, the local authority will pursue the tenant, who is liable by statute, despite his or her agreement with the landlord. (Of course, the tenant may then pursue the landlord for reimbursement.)

Liability of non-resident owners of occupied property

There are a few less common situations in which the owner is in fact liable for an occupied home under the Council Tax (Liability for Owners) Regulations 1992 SI 1992 No 551 (as amended by the SI 2000 No 537) (see Chapter 6). Thus, owners of prescribed classes of occupied dwelling have liability for council tax. These are

- Class A — residential care homes
- Class B — religious communities
- Class C — houses in multiple occupation
- Class D — resident staff
- Class E — ministers of religion
- Class F — accommodation for asylum seekers.

Council tax liability for an occupied residence will only fall to a non-resident owner if the property fulfils one of the statutory definitions relating to the above six classes. "Owner" in this context means non-resident leaseholder with a leasehold interest of at least six months or if there is no such person, then the non-resident freeholder. Box 12.3 shows a summary of the main defining characteristics of the six classes of property for the owner's liability (see also *Unoccupied property* on p139).

Class C: Houses in multiple occupation

As seen above, the local authority can only hold a non-resident landlord of an occupied home liable, if it falls into one of the above classes. Most of these will not be encountered by a typical landlord, accept for Class C, houses in multiple occupation (HMO). The definition was broadened by the Council Tax (Liability for Owners and Additional Provisions for Discount Disregards) (Amendment) Regulations 1995 SI 1995 No 620 in 1995 to include homes that met either, rather than both of the criteria mentioned in Box 12.3. Landlords should take care, because a dwelling can be retrospectively designated as an HMO. The landlord will be the person held responsible for ensuring that council tax is paid on the HMO, even if there is a contractual term providing that the tenant or licensee shall pay. The *Valuation Tribunal Service LPAC Newsletter*, October 2004, gave an account of a case involving retrospective backdating taken by the West Midlands (West) VT. The tenancy agreement for the whole house was complicated since the rent was equivalent to a rent for a one bedroom flat and the tenant (an employee of the landlord's company) was restricted in the use of the accommodation — one room was an office and the landlord occupied two other rooms when visiting the town (about five months

---

Box 12.3 Occupied homes giving rise to the non-resident owner's liability

Class A	Residential care home	• nursing home or mental nursing home within the meaning of the Registered Homes Act 1984 and certain similar properties • care home within the provisions of the Care Standards Act 2000 • approved bail or probation hostel • other homes used at least mainly for persons who require personal care by reason of old age, disablement, alcohol or drug dependence or mental disorder • residential accommodation provided under section 21 of the National Assistance Act 1948 • Abbeyfield Society home
Class B	Religious community	• dwelling inhabited by a religious community whose principal occupation consists of prayer, contemplation, education, the relief of suffering, or any combination of these
Class C	House in multiple occupation	• dwelling constructed or adapted for occupation by persons who do not constitute a single household or • dwelling where the occupier or occupiers each have a contract or licence to occupy only part of the dwelling
Class D	Resident staff and family	• dwelling occupied by domestic servant or servants and their family, which is occasionally occupied by the owner
Class E	Ministers of religion	• a dwelling which is inhabited by a minister of any religious denomination as a residence from which he performs the duties of his office
Class F	Asylum seekers	• dwelling provided to an asylum seeker section 95 of the Immigration and Asylum Act 1999

---

each year). It was held that the property was a HMO and the council tax should be back-dated (see Chapter 8).

Where the dwelling is not an HMO, the landlord may still have a type of obligation to pay council tax. For example, where the lease or tenancy provides that council tax is payable with the rent, as discussed, this obligation is by way of contract only and is only enforceable by the tenant. The tenant remains by statute liable to ensure that the council tax is paid. To ensure enforceability of the bill, revenue practitioners should ensure that the tenant is named on the bill if the dwelling is not a HMO. However, in the interest of practical ease for all parties, it is not uncommon practice for a local authority to bill a reliable landlord who is in the habit of collecting council tax inclusive with rent and paying it over to the local authority.

Class E: Ministers of religion

The Department for the Environment, Food and Rural Affairs has provided guidance on the activities of ministers of religion.

Unoccupied property

Owners (or tenants) of unoccupied property

An owner occupier or a landlord who has an unoccupied or vacant dwelling will be responsible for any liability that arises for the dwelling (unless an absent tenant is liable because he or she has a current lease that has not been surrendered). There are discounts and exemptions that apply to uninhabited dwellings for which purpose a distinction is made between "unoccupied" meaning furnished, and "vacant" meaning unfurnished.

On becoming unoccupied a dwelling will immediately attract a discount of between 10% and 50%. The amount will depend on the policy of the local authority that has the discretionary power to vary the discount between these two extremes under section 75 of the Local Government Act 2003. The Council Tax (Prescribed Classes of Dwellings) (England) Regulations 2003 SI 2003 No 3011 provides more detail of this. The discount awarded will continue for the unoccupied dwelling until it its status changes, ie becomes vacant, occupied, uninhabitable, demolished, or turned into something that is no longer a dwelling.

Owners of vacant dwellings

On becoming vacant a dwelling will enjoy a 100% exemption from council tax for six months as provided by Class C of SI 1992 No 558 as amended by SI 1993 No 150, after which from 50% to 100% of the usual full tax will be payable under section 11 of the 1992 Act. The precise amount will depend on the discretionary policy of the billing authority.

On completion, a newly built vacant property will enjoy a period of six months exemption, after which the owner becomes liable to council tax of from 50% to 100% of the full liability (under Class A of SI 1992 No.558 as amended by SI 2000 No 424). If a newly completed property is being used as a show home, then business rate is likely to be payable instead of council tax.

Owners and occupiers of caravans

Depending on the circumstances, either the owner or the occupier of a caravan may be liable to council tax; but sometimes no local tax applies or there may be a liability to business rates. Thus, the treatment of the owners and occupiers of caravans receive depends on the type of occupation or use of the caravan. For council tax purposes a key issue is whether the caravan is a person's sole or main residence. In essence there are several types of case as shown in Box 12.4.

A similar analysis applies to boats. The Valuation Office Agency document *Practice Note 7: Application of council tax to caravan pitches and moorings* gives a detailed account of the historic case law and the issues to be considered.

Issues concerning, for example, caravans and agricultural dwellings which are subject to planning restrictions or conditions are considered in Chapter 16 (p185) from a valuation perspective. (The confusion concerning these conditions or restrictions is also touched upon there, but see p146.)

Box 12.4 Treatment of the owners and occupiers of caravans

Owner occupation
- sole or main residence
- not the sole or main residence

- owner occupier is liable as if a normal dwelling
- treated as a second home if it is on its own site

Tenant in occupation
- the tenant occupies it permanently, ie sole or main residence

- tenant is normally liable for council tax as a resident

Householder's holiday caravan
- parked at home
- touring caravan — not let
- let for short periods, being taken from home
- on site, let to holidaymakers

- owner not liable
- no liability
- no liability (may be income tax)
- business rates apply

Caravan site operator
- on-site caravans

- business rates apply

Vacant caravan pitch

- not liable under Class R of SI 1992 No 558 (as amended by SI 1994 No 839)

Composite property — residential occupiers

Where a factory or shop, for instance, contains residential accommodation for the owner or staff, an assessment for council tax would have been made and liability arises on the occupier. The employer may have contracted to pay the council tax on the employee's behalf. Where this happens the employee may be liable for income tax on deemed income (the amount of council tax) but there are exceptions.

It may be noted that special assessments or valuations may be needed to ascertain the banding for such residential accommodation (see Chapter 16).

Exemptions, Reliefs, Discounts and Reductions

Aim

To identify and explain the numerous exemptions, reliefs and reductions from council tax

Objectives

- **to identify those who are exempt from council tax**
- **to identify those who enjoy reduction or relief from council tax**
- **to explain the detail of the principal exemptions and reliefs on individuals**
- **to identify and explain the dwellings exempt from council tax**

Introduction

A number of part or full exemptions, reliefs, discounts, and reductions (and instances of non-taxation) are available to those who would otherwise pay council tax. They may be grouped as follows:

Non-Taxation

- non-taxation for some adults because they are not liable.

Discounts

- discount arising because, after disregarding any persons listed in Box 13.1 (the so-called "discount disregards") there is only one person deemed as a resident
- discount arising because a dwelling is unoccupied
- discount arising because a dwelling is vacant.

Exemptions

- total exemption of unoccupied dwelling because of the status of the absentee

- total exemption of unoccupied or vacant dwelling because of the status of the dwelling
- total exemption of unoccupied or vacant dwellings because of the status of the owner
- total exemption of occupied dwelling because of the status of the occupiers.

Other reductions

- reduction arising from the disability of an occupier
- pensioners council tax assistance
- other reductions for any individual or group at the billing authorities discretion
- transitional relief after revaluation
- contractual imposition of the burden of the council tax to a third party
- special reductions for lump sum payment, or payment by direct debit
- means tested council tax benefit.

This chapter examines each of the above except the following:

- council tax benefit (see Chapter 9)
- reductions for lump sum payment and payment by direct debit (see Chapter 8)
- the exclusion of Crown exemption.

Non-taxation

Council tax liability applies to those who fall to be liable according to statutory conditions. These are shown in detail in Chapter 12. All other adults despite sometimes having gainful employment or significant financial means effectively enjoy immunity from the tax, for example:

- adults living with liable parents (so long as they are not sharing the tenancy, freehold, or leasehold with their liable parents)
- anyone who lives in a dwelling with another occupier that has a superior material interest in the property (so long as its not their spouse), for example, a lodger
- relatives or friends living permanently at the address (having no sole or main residence elsewhere, and not sharing the tenancy freehold or leasehold with the liable occupiers).

In such circumstances there is arguably a measure of unfairness in the council tax regime and this is often pointed out as the disadvantage of the property based council tax by those who support alternatives such as a poll tax or a local income tax.

Discounts

Under section 11 of the Local Government Finance Act 1992, an occupier who lives alone or singly and is aged 18 years or more will be eligible for a 25% discount from the full council tax. Also, although an individual may occupy a building with others, he or she may still be eligible to the 25% discount as a lone occupier if the other person or persons are not counted as residents for council tax purposes. Such persons are known as discount disregards and are shown in Box 13.1. They are eligible to be

---

Box 13.1 Persons not counted when allowing the single occupier discount

Individuals
- whose main residence is elsewhere
- under the age of 18 years

Young adults and children for whom child benefit is paid, where they are :
- in their 19th year

Students
- under 20 years of age who have recently left education after 30 April will be discounted until 1 November of the same year
- under 20 years studying A levels, Scottish highers or other equivalent
- in full-time education, ie in university or college

Young work students or trainees, such as:
- apprentices
- student nurses
- language assistants

Non-permanent residents, eg guests on holiday with the taxpayer

Patients and those in care, namely:
- those in hospital
- those in hostels, nursing homes and residential care homes with a high degree of care

Residents of no fixed abode staying in hostels, night shelters and the like (applies to England and Wales)

Certain international visitors:
- those of armed forces
- those of foreign embassies and the like
- those of international defence bodies

Detainees
- in prison
- in other forms of detention

Persons who are severely mentally impaired

Care workers living in at the dwelling and giving care there

Member of religious community, ie one for prayer, education, and the relief of suffering, where the individual:
- has no income or capital (disregarding any pension)
- is dependent on the community to provide for his or her material needs

---

disregarded by virtue of the Council Tax (Discount Disregards) Order 1992 SI 1992 No 548 (as amended) and the Council Tax (Additional Provisions for Discount Disregards) Regulations 1992 SI 1992 No 552 (as amended).

Where all the occupants qualify to be disregarded, the dwelling usually falls into an exempt category, see exemptions classes M, N, O, P, S, U and V under exemptions described below; otherwise a 50% discount applies.

Discount for an unoccupied dwelling

An unoccupied dwelling, ie one that has at least some furniture other than carpets and curtains, but is nobody's sole or main residence, attracts a discount of from 10% to 50%. The precise percentage for a given dwelling is at the discretionary policy of the billing authority under which the property is situated.

Discount for a vacant dwelling

A vacant dwelling, ie one that is nobody's sole or main residence and is substantially unfurnished, will first be subject to a total exemption for six months (see Class C under exemptions later in this chapter) before any discount becomes relevant. After the expiry of the six months' exemption, the vacant dwelling will give rise to a discount from nil to 50%. The precise percentage for a given dwelling is at the discretionary policy of the billing authority under which the property is situated

Exempt dwellings

Exemptions are total, in that they give rise to a nil council tax liability, however depending on circumstances they are not always for an indefinite period.

There are 23 prescribed classes of dwellings that are exempt under section 4 of the 1992 Act. They are classified in article 3 of the Council Tax (Exempt Dwellings) Order 1992 SI 1992 No 558, (as amended) which lists the exemption classes from A to W. The alphabetical order has been forgone in Box 13.2, so that the exemptions can be grouped into four categories as follows:

- total exemption of unoccupied dwelling because of the status of the absentee
- total exemption of unoccupied or vacant dwelling because of the status of the dwelling
- total exemption of unoccupied or vacant dwellings because of the status of the owner
- total exemption of occupied dwelling because of the status of the occupiers or usual occupiers

Further details of each exemption class are given below.

Class A: Major Repair

This exemption applies to an unoccupied and unfurnished dwelling if the property:

- needs major repair work to make it habitable or
- is undergoing major repair work to make it habitable or
- is undergoing structural alteration which is not yet substantially complete or
- has been continuously unoccupied and unfurnished for a period of less than six months following the completion of the above major repair work, or structural alteration
- exempt class A applies for a maximum continuous period of 12 months.

Class B: Charity

This exemption applies if the property:

Box 13.2 Dwellings which are exempt from council tax

Class Exempt because of the status of the absentee

D   Unoccupied property — vacated by a liable person being held in detention.
E   Unoccupied property — vacated liable person living in a hospital or a care home
I   Unoccupied property — vacated by a liable person living elsewhere to receive care
J   Unoccupied property — vacated by liable person living elsewhere to provide care to someone else.
K   Unoccupied property — vacated by a liable person living elsewhere to attend college or university
H   Unoccupied property which is awaiting occupation by a minister of religion

      Exempt because of the status of the dwelling

A   Property undergoing structural repair (6 or 12 months)
C   Empty and unfurnished property (6 months)
G   Unoccupied property which is prohibited from being occupied by law
R   Unoccupied caravan pitches or boat moorings
T   Unoccupied annexes that are restricted from being let separately by planning controls

      Exempt because of the status of the owner

B   Empty property owned by a charity (up to six months)
F   Unoccupied property where no probate or letters of administration have been granted (up to six months after the grant of probate or letters of administration)
L   Unoccupied property which has been repossessed by a bank or a building society
Q   Property which comes under the responsibility of a bankrupt's trustee

      Exempt because of the status of the occupier(s)

M   Halls of residence.
N   Property which is occupied only by students
S   Property occupied only by people under the age of 18 years
U   Property occupied only by people who are severely mentally impaired
V   Property occupied by diplomats
O   Property used as armed forces accommodation
P   Property which is used as visiting armed forces accommodation
W   Property which is a separate annex and is occupied only by a dependent relative

---

- has been unoccupied (but not necessarily unfurnished) for less than six months
- is owned by a body established for the purposes of charity only
- was last occupied in furtherance of charitable purpose.

(The Charity Bill which is likely to receive royal assent in late 2005 provides for the extension of the meaning of "objects" which are considered charitable within the provisions of the Charity Acts.)

Class C: Vacant

This exemption applies if the property:

- has been unoccupied and unfurnished for less than six months.

Billing authorities have recently been given discretionary powers to partially or wholly remove the 50% discount available for unoccupied second homes. If such a property is unfurnished it will be exempt under class C for the first six months.

Class D: Detention

This exemption applies if the property:

- is unoccupied because the owner or tenant who would occupy it as their sole or main residence, or who last occupied it as such is detained in prison or some other type of formal detention. This exemption does not apply if the person is in prison for non payment of council tax or fines.

Class E: Away in a nursing home

This exemption applies if the property:

- is unoccupied because the owner's or tenant's sole or main residence has changed from the property to a nursing home, residential care home, hospital private hospital or other such qualifying home providing personal care and support.

Class F: Awaiting probate

This exemption applies if the property:

- has been unoccupied since the death of a person who had a freehold or leasehold interest in the property and six months has not yet expired since the grant of probate or letters of administration. (In the case of a deceased tenant this exemption applies for up to the same duration, only if the executor or administrator is liable for rent or a licence fee on the property.)

Class G: Prohibited by law

A dwelling is exempt when it is unoccupied under action to keep it empty under powers conferred by any Act of Parliament prohibiting property being kept empty as a result of, for example:

- an intended purchase under an Act, eg a compulsory purchase order
- a closing order or a demolition order.

As there is some confusion on this matter the ODPM is or is about to consult on a proposed amendment to Class G to clarify the exemption to dwellings subject to a planning restriction to occupancy (see Council Tax Information Letter 5/2005). Such conditions are briefly dealt with in Chapter 16 (see p185).

Class H

This exemption applies if the property:

- is unoccupied and is being held for the purpose of occupation by a minister of any religious denomination for the purpose of carrying out his duties.

Class I: *Away receiving care*

This exemption applies if the property:

- is unoccupied because the owner or tenant who previously occupied it now has their sole or main residence elsewhere for the purpose of receiving care for old age, disablement, illness, past or present alcohol or drug dependence or past or present mental disorder, but does not otherwise qualify under class E.

Class J: *Away providing care*

This exemption applies if the property:

- is unoccupied because the owner or tenant who previously occupied it has moved their sole or main residence to another place so that they can provide, or better provide, personal care for a person who requires such care because of old age, disablement, illness, past or present alcohol or drug dependence or past or present mental disorder.

Class K: *Left unoccupied by a student*

This exemption applies if the property:

- is unoccupied and the otherwise liable party is a student or has become a student within six weeks of leaving the property. The definition of student for this purpose is complex but generally includes a requirement that they must be studying in a recognised educational establishment for at least 21 hours per week and for at least 24 weeks of the year.

Class L: *Mortgagee in possession*

An owner of dwelling may borrow money and offer the property as security for the loan. The parties usually enter into a mortgage. If the borrower subsequently defaults, one of the remedies available to the mortgagee is to take possession of the property in accord with the terms of the mortgage contract, ie with the view to its disposal. Provided the property is not occupied, the mortgagee in possession is not liable to pay council tax. (The property is exempt under section 4 of the 1992 Act.)

Class M: *Student halls of residence*

This exemption applies if the property:

- is a student hall of residence

The exemption will also subsist during periods of vacation for any hall of residence.

Class N: Occupied by students

This exemption applies if the property:

- is occupied only by students either as a main home or residence or as term time accommodation or
- is occupied as described above and is also occupied by a student's spouse who is not a British citizen and who is prevented from claiming benefit or taking paid employment by the terms of the permission to stay in the country (see SI 1995 No 620 which extended SI 1992 No 558).

For this kind of accommodation, the exemption continues provided the student or students will return for the next term's course.

The Valuation Tribunal Service LPAC Newsletter of October 2004 gave a long and interesting account of at least three valuation tribunals which have regarded nannies or housekeepers employed by foreign students as "relevant persons", ie as dependants within the meaning of SI 1992 No 558, as amended. While in the country the employees could not work for other employers or claim benefits, but the SIs contain no definition of "dependant" — so dictionary definitions (which included "servants") were used as a basis for the valuation tribunal's decisions.

Class O: Armed forces

This exemption applies if the property:

- is owned by the Secretary of State for Defence and is held for the purpose of occupying armed forces.

Chapter 12 deals with aspects of the exclusion Crown exemptions in respect of those held for Crown purposes by certain bodies (see p135).

Class P: Visiting armed forces

This exemption applies if the property:

- at least one of the occupants is a member of qualifying visiting armed forces and is part of a visiting detachment of those forces.

Class Q: Trustee in bankruptcy

This exemption applies if the property:

- is held by a trustee in bankruptcy.

Class R: *Pitch or mooring*

This exemption applies if the property:

* is a pitch or mooring that is not currently occupied by a caravan or boat.

Class S: *Occupiers under 18 years' of age*

This exemption applies if the property:

* is occupied only by persons under 18 years' of age.

Class T: *Annex prohibited from letting*

This exemption applies if the property:

* is an unoccupied annex or similar self-contained adjoining property that is prohibited from being let separately under section 171A of the Town and Country Planning Act 1990.

Class U: *Occupied by persons severely mentally impaired*

This exemption applies if the property:

* if the property is occupied only by one or more severely mentally impaired persons. The definition of severely mentally impaired includes a requirement for certain benefits to be received and for a doctors to designate the person as such.

Class V: *Diplomats*

This exemption applies if the property:

* in which at least one of the occupants benefits from qualifying diplomatic immunities and privileges.

Class W: *Annex occupied by dependent relative*

This exemption applies if the property:

* is an annex, eg so-called granny flat, which is occupied by a dependent relative of the person who has their sole or main residence at the dwelling to which the property in question is annexed.

Individuals with disabilities

There are three forms of relief which recognise that dwellings in which individuals with mental or physical disabilities reside have been adapted to meet their requirements, namely:

Box 13.3 Common adaptations which do not in themselves qualify

Deafness	• hearing loop in a room • doorbell which flashes • telephone which flashes • vibrating pad linked to smoke alarm system • communication equipment, eg telephone, facsimile, personal computer and ancillary equipment
Blindness	• tape recorder • communication equipment (as above) • Braille equipment • sound and vibrating pad linked to smoke detection system • a specially fitted kitchen
Limited or total immobility	• a specially fitted kitchen • grab rails in a bathroom • special ablutions or bathing facilities, eg bath with a door or hoisting equipment
Special requirements	• special lavatory fitments, low sink • kidney problems — dialysis machine

---

- a valuation reduction which is sufficient to reduce the valuation band applied to the property (see Chapter 15)
- an exempt dwelling by those with severe mental impairment (see Class U p149)
- the equivalent of a band reduction granted because of an adaptation to a property that is to accommodate the person with a disability (see Council Tax (Reductions for Disabilities) Regulations 1992 SI 1992 No 554, as amended).

Adaptations to property

An individual with a disability may be able to obtain some relief from council tax for adaptation works to the property. However, the differing ways in which the provisions are interpreted and applied appear to cause happiness or distress. Box 13.3 shows some of the adaptations which may be made to dwellings to meet the requirements of a person with one or more disabilities but are unlikely to qualify.

It does not follow that a council tax relief will be afforded as a result of a particular adaptation — the wording of SI 1992 No 554 and its interpretation is all important. In essence, the wording appears to exclude many adaptations, eg a night storage heater (see *Williams (Howell) v Wirral Borough Council* [1981] RA 189 and a bathroom conversion (see *Luton Borough Council v Ball* [2001] EWCH Admin 328).

The measure of relief is to remove the dwelling from its original band to the next below it, eg a band C dwelling drops to band B. In the case of a dwelling in band A, one sixth is taken off (this is provided for in the Council Tax (Reductions for Disabilities) (Amendment) Regulations 1999 SI 1999 No 1004).

In fact it is unlikely that the items in Box 123.3 will qualify but may do so in a larger scheme of adaptations which meets the requirements of the statutory provisions. They allow a band reduction on application if a person (adult or child) who is mainly resident at the dwelling is substantially and permanently disabled and the property has at least one of the following:

- a room which is not a bathroom, a kitchen or a lavatory — which is predominantly used by and is essential or of major importance to the well-being of the disabled resident by reason of the nature and extent of his or her disability or
- a bathroom or kitchen which is not the only bathroom or kitchen within the dwelling and which is required for the needs of any qualifying individual resident in the dwelling or
- sufficient floor space to permit the use of a wheelchair required for meeting the needs of any qualifying individual resident in the dwelling. (The wheelchair must be relied upon by the disabled person to get around inside the dwelling).

Transitional relief after revaluation

When a new valuation list comes into force following a revaluation, most taxpayers who face a large increase in council tax, will enjoy transitional relief. The exceptions are a new owner of any dwelling.

Chapter 14 gives more detail of the nature of the relief and how it operates in Wales from the 1 April 2005 (see p161).

Other powers for relief

Pensioner's council tax assistance

From time to time the government, in its budget, announces help for pensioners with their council tax. This is usually paid out alongside winter fuel payments and, therefore, the billing authority plays little or no part in its administration or finance.

For 2005–2006 £200 was given to pensioners paying council tax in the year. The payment will not be available to pensioners who receive the pension credit guarantee since they receive council tax benefit already. These payments are related to the tax-free payments of £100 or £50 for those who are 70 years or over (see the Age-Related Payments Act 2004).

Power to reduce council tax

Billing authorities have been given discretionary powers by section 76 of the Local Government Act 2003 to partially or wholly reduce the council tax due in any particular case or class of case as it thinks fit. For example, a few authorities have used this power to reduce the council tax burden on pensioners.

Contractual imposition of the burden of the council tax to a third party

The most common example of this, is where a liable tenant does not need to pay because his landlord has accepted the burden of the council tax within the terms of the tenancy agreement. However, on non payment the billing authority will correctly enforce collection from the liable tenant. The tenant may in turn be able to recover his losses from the landlord by enforcing the terms of his tenancy contract.

Part 6

Valuation Lists and Assessment

Compilation and Maintenance of the Valuation List

Aim

To examine the compilation and maintenance of the valuation lists in the United Kingdom

Objectives

- to explain the need for revaluations
- to describe the procedures for a typical original list and its revaluation
- to describe the transitional arrangements in Wales

Introduction

The original valuations lists were compiled for 1 April 1993 at values of 1 April 1991. However, by virtue of section 3 of the Local Government Finance and Valuation Act 1991 all dwellings in Great Britain were originally assessed for council tax so as to place each in a valuation band at the appropriate time for each country. It is emphasised that the intent of the legislation is to allocate a dwelling to a band of value rather than to obtain a precise value for it. For instance, many properties were valued from the "roadside" for the original valuation list.

By virtue of the 1992 Act, for each billing authority area, a listing officer (assessor in Scotland) was subsequently responsible for:

- compiling the valuation list for the area (essentially from the values previously obtained)
- maintaining the valuation list once it comes into force.

He or she is appointed to the Valuation Officer Agency (VOA) (assessors were directed) by the Commissioners of Inland Revenue (now within HM Revenue & Customs) under section 20 of the 1992 Act.

This chapter examines how a valuation list is created and then renewed by revaluation. The maintenance of the valuation list and the statutory basis and approaches to valuation for council tax and domestic rating, the latter in Northern Ireland, are dealt with in Chapter 15. Also, practical aspects of valuation are covered in Chapter 16.

Matters of dispute arising from revaluation work by the VOA are deal with by the valuation tribunals, courts, internal complaints procedures or the ombudsmen in Chapters 17 and 18.

Valuation list — contents

Section 23 of the 1992 Act gives, and the Council Tax (Contents of Valuation Lists) Regulations 1992 SI 1992 No 553 prescribes the information required in a valuation list. The billing authorities' original valuation lists for England, Scotland and Wales contained details of the chargeable dwellings for each authority's respective area. Each entry was required to give the following information:

- a reference number given by the listing officer
- the address of the dwelling
- the valuation band of the dwelling
- for composite property a notation — "Comp".

Need for revaluations

Generally, the legislation has a built in expectation that there is fairness. One way of achieving this would be to have regular "revaluations", ie a fresh valuation of every property in the country. Although the 1992 Act did not provide for it, the expectation was that revaluations should take place every few years but in practice successive governments have allowed slippage in any notional programme for council tax. Thus, as time passes property values change and the valuation list will have become out of date. Any two of the properties shown on the valuation list may be similar original values (albeit within a band and not shown on the valuation list) but the current open market values of the two dwellings may have become markedly different as the years have passed. It is conceivable, if not certain, that the occupiers paying similar levels of tax should not be treated in the same way — but this is not the case while the original list remains in force. The principle of "fairness" has been breached. It is important, therefore, for regular revaluations to take place. (This assumes that council tax is intrinsically fair — a debateable issue!)

Revaluation cycle — fairness

Although non-domestic rating revaluations have been carried out to a five yearly cycle in recent years this has not been the case for council tax. It has been be argued that regular revaluations enhance the principle of fairness in local taxation — in *Strong Local Leadership: Quality Public Services* (a White Paper 2001) — the government proposed a 10 year cycle for council tax revaluations. As a result section 77 of the Local Government Act 2003 provided that after the 2005 revaluation in Wales and the 2007 revaluation in England (now postponed), they should take place every 10 years in each country respectively. It remains to be seen whether a period of 10 years is a sufficiently short for unfairness to be removed. The intended effect is to enable continued accuracy in relative values of each property. (However, draft legislation amends section 77 of the 2003 Act — removing the 10-year cycle.)

Market movements

It should be borne in mind that there is a period of two years from the date when all properties are valued (the antecedent date) and the date when the valuation list comes into force. From the 1 April 1991 there was a marked decline in overall market and the decline varied for some areas within the same billing authority's area. The result was that some taxpayers with properties of similar values on the antecedent date found that the same properties had a value differential two years later. For 2005 the growth in house prices is slowing again and the same issue may arise by 1 April 2007. As yet, the government does not appear to have addressed the matter.

Current cycle of revaluations

Revaluations are being or have been carried out in the United Kingdom. They cover two of the four countries separately. Those in Northern Ireland and Wales were completed in 2001 and 2005, respectively — although the former has a different basis of assessment.

Such revaluations do not in themselves give rise to any increase in the aggregate of council tax payable. However, such an increase may be experienced where the capital value of a residence or group of residences has, for any reason, increased disproportionately to other values in a billing authority's area. There are two effects which may cause this:

- individual taxpayers may suffer increases due to movement upwards within the structure of bands
- as between billing authority areas, the government's grant system may be altered to the detriment of some of them.

Wales — revaluation

The Valuation Office Agency's council tax revaluation for Wales has just been completed. The new list operates from 1 April 2005 when the billing authorities will charge council tax payers. Of course occupiers, having already received notification of the new banding, needed to consider whether they need to take action, including:

- appealing against inappropriate banding, ie from 1 April 2005
- whether financial assistance is needed, ie making an application for council tax benefit
- where any rent is inclusive, is there scope to increase it if the council tax increases.

Where an increase in banding will operate owners will normally enjoy transitional relief.

England — revaluation

Having completed the so-called "Council Tax Reval 2005" for Wales, the VOA had turned its attention to England's list for a revaluation but this has been postponed. The date of valuation should have been 1 April 2005 (see paragraph 44 of schedule 7 to the Local Government Act 2003) and the new lists should have operated from 1 April 2007, ie bills based on the new valuations (and bands, it seems) should have been issued for 1 April 2007— they would have contained just over 22 million dwellings.

The revaluation would have been based on property values at 1 April 2005. It may be of interest to note that from April 2005 computer modelling would have been used in the work for the revaluation in England. A contract has been placed with Cole Layer Trumble Company (USA) to supply the system which has already faced trials. For the revaluation, local market data and information about the individual properties in an area will be used to provide the banding for the local list.

Northern Ireland — revaluation

The 700,000 dwellings in Northern Ireland have been revalued to capital values but the results are used in the traditional rating approach to local taxation adopted there (council tax is not used in Northern Ireland). The assessments are not banded, so an individual property has its own assessment for the purposes of charging rates (see Chapter 1, p14).

Scotland — revaluation

Scotland has had no revaluation since the original council tax revaluation. Normally, it might be expected that a revaluation would occur within the cycle for revaluations in the United Kingdom. However, a Bill was progressing through the Scottish Parliament to change local taxation in Scotland by abolishing council tax and introducing "service tax". It was proposed that the new tax will become effective in 2006 but the Bill was not passed into law.

Procedures for revaluation

Section 22B(7) of the 1992 Act provides for the listing officer to assess every chargeable dwelling and to prepare the valuation list for the due date. For authorities in England, the listing officer — sometimes referred to as the "valuation officer" or less accurately the "district valuer" — had the duty to compile the next valuation list on 1 April 2007. However, the draft should have been sent to the authority sometime before 1 September in 2006 and the completed list, ie on the information then held deposited by 1 September 2006. Box 14.1 summarises the situation for Wales regarding the dates for the revaluations, giving the originally intended dates for England.

Box 14.1 Summary of the timetables for the revaluations in Wales and England

Wales	Date of valuation	1 April 2003
	Date for draft submission	1 September 2004
	Date of compilation	1 April 2005
	Dates for transitional arrangements	1 April 2005 to 1 April 2008, ie three years
England	Date of valuation	1 April 2005*
	Date for draft submission	1 September 2006*
	Date of compilation	1 April 2007*
	Dates for transitional arrangements	not yet available

Note: *now postponed

Private sector involvement

Although the valuation of dwellings for the first valuation lists for Great Britain was the responsibility of government officials much of the work was undertaken by 337 firms in the private sector by virtue their appointment under section 3(4) of the Local Government Finance and Valuation Act 1991 (now repealed). They worked, of course, to the statutory definitions and assumptions (see p169) — the date of valuation was 1 April 1991. Each potential contractor was required to register and complete a pre-qualification form for the work which was to be undertaken in the period from 10 September 1991 to 27 September 1991. The pre-qualification criteria related to the following:

- a firm's professional competence
- capacity to do the work
- financial viability to do the work.

Valuers tendered for work which ranged from about 8,000 to 20,000 dwellings; receiving weekly batches of properties over several weeks from about the turn of the year in 1991–1992. Those who were appointed did not have access to the inside of the dwellings when valuing them but did have information as follows;

- that collected under sections 4 and 6 of the 1991 Act
- the VOA's 1973 database
- in-house records.

Unwarranted use or disclosure of official information is an offence in England and Wales and in Scotland under sections 3(6) and 5(6) of the 1991 Act respectively. Other information may have been protected by the data protection legislation.

It is not clear whether a similar arrangement would have been adopted for the England's council tax revaluation which was due to begin in 2005.

Computer modelling

It was intended that from April 2005 computer modelling would be used in the work for the revaluation in England. A contract has been placed with Cole Layer Trumble Company (of the USA) to supply the system which has already faced trials. For the revaluation, local market data and information about the individual properties in an area may be used to provide the banding for the local list in the future.

Accuracy

The Keen and Travers study (1993) and other work touched up on accuracy in the valuation work. It seems that the original valuation lists had a tendency to "built-in" inaccuracies which were caused by the likes of:

- "dwellings" being wrongly identified as such
- new dwellings being left off
- dwellings being located in the areas of other billing authorities

- the valuations being incorrectly performed
- the identified anomalies were not being corrected before the list had become operational.

Much of the inaccuracy may have been due to the database being somewhat inadequate — it was an updated from:

- the old domestic rating list
- the community charge database
- such records as might have been available, eg stamp duty returns.

At present the database is much more comprehensive and becoming increasingly so with such initiatives as:

- the National Spatial Address Infrastructure
- the Valuebill Project.

The latter is a VOA pilot project to create a common property database for participating billing authorities and the VOA. The original number of participants has already been increased.

The private sector involvement was wide and some inaccuracy may have crept into the valuation work. However, of the 350 or so contracts only a handful were rescinded by the VOA. Also, the work was monitored and some results were checked in detail — any found to be inadequate were repeated. Nevertheless, it was thought that about 10% of the entries were inaccurate.

Appeals

Appeals following revaluation are made to the valuation tribunal. They arise for the following kinds of reasons:

- there has been a mistake, eg the property is not as described, no longer exists or is not in the billing area (but see *Property straddling borders* in Chapter 16)
- the property is considered as being over-valued — it is in a band which is too high.

After the original valuation lists came into force on 1 April 1993 there were about three-quarters of a million appeals 1991 made in the six month period allowed. It was estimated that 10% of the assessments were in error but in some areas with a high number of relatively low-banded dwelling the rate of appeal was about 1.5%; in some areas with a high number of high-banded dwellings there were about 4.5% of appeals.

The appeal process is relatively straight forward. In many instances, the taxpayer may conduct the appeal alone, particularly where a mistake has occurred. In other cases it may be worthwhile having a professional valuation undertaken and representation. Again, if the taxpayer obtains sufficient information of transactions at about the time of the antecedent date, the band allocation may reflect a glaringly obvious over-valuation. However, in the case of an unusual property or a complicated case, eg a composite property, it may be better for the taxpayer to have a professional valuer's report to support negotiations with the listing officer or to have before the valuation tribunal on appeal.

An appeal to the High Court may be made under Order 55 of the Rules of the Supreme Court 1965 SI 1965 No 1776 on a point of law as a result of the Council Tax (Alteration of Lists and Appeals) Regulations 1993 SI 1993 No 290 (see regulation 7).

A judicial review may be sought if the decision of the valuation tribunal is considered to be based on one of the following:

- illegal
- irrational
- based on a procedural fault.

Appeals Direct

The prospect of revision of the appeals procedure is being investigated — it is called "Appeals Direct" which should be in place for the next revaluation of council tax in England. Legislation provides that appeals against a revaluation determination must be made by 30 November of the valuation list coming into force, ie from 1 April.

Changes to the structure of bands

The original valuation lists were based on bands A to H. However, section 78 of the Local Government Act 2003 introduced the power to alter the number of council tax bands (by insertion of subsection (4A) into section 5 of the 1992 Act). The revaluation for England was expected to be accompanied by the addition of extra bands but it seems that the government has decided not to introduce any change for England. This was expected to redistribute the burden of council tax and it was expected that the extra bands would have mostly impacted on residences of the highest values. The new band structures in Wales have had an impact on the way in which the transitional arrangements will operate in Wales (see below).

The regressive lack of fairness of the banding system is commented on elsewhere — Wales has had the number of bands increased by one band (band I); it has been suggested that the number of bands in England should be nearly doubled (but it seems doubtful that this will happen).

Transitional relief after revaluation

Following a revaluation, any allocation of the dwelling into a higher band would normally result in an increase in the council tax payable when the new valuation list comes into force on the 1 April of the relevant year, eg 1 April 2005 for Wales. However, after each revaluation it is accompanied by the introduction of transitional provisions under section 13B of the 1992 Act. The regulations phase in the impact of any significant increase or decrease in amount payable on any individual property or group of properties.

The National Assembly for Wales has made regulations for arrangements to smooth changes in council tax in Wales as a result of the new revaluation. (Similarly, the secretary of state would have done the same for England.) In Wales the transition period will last for three years from the year 2005–2006 to the year 2007–2008. A taxpayer will be eligible for relief provided the criteria summarised in Box 14.2 are satisfied. As a result of the Council Tax (Valuation Bands) (Wales) Order 2003 SI 2003 No 3046 (W271) the number of bands in Wales has been increased by adding band I (see Chapter1). It may be noted that the last couple of bands are about £100,000 in depth so that those with dwellings which are exceedingly high in value, say over £600,000 to several million pounds, still pay the same amount as those whose property is marginally over £525,000.

Box 14.2 Criteria for qualifying for transitional relief in Wales

Valuation list inclusion	• the dwelling must be included on or for the valuation list for 31 March 2005 (the old list)
Increase in bands	• the dwelling must have increased in value by at least two bands for the 1 April 2005 list (new) compared with the 31 March 2005 list (old)
The taxpayer(s) remain	• the person liable on 31 March 2005 (old list) is also liable on 1 April 2005 (new list) and throughout the three years

---

The relief is automatically given where the re-banding will result in a jump in liability of two or more bands. The measure of relief is not be more than one band, two bands and three bands increase in each of the three years successively.

Each year, the billing authorities will be able to claim from the National Assembly for Wales the council tax foregone as a result of the transitional relief, subject to certain discounts and other matters not covered by the relief.

New valuation lists for England

On 13 October 2006 the Council Tax (New Valuation Lists for England) Bill was introduced in the House of Commons. It provides for amendment by part repeal of section 22B of the 1999 Act (which came from section 77 of the Local Government Act 2003). The Bill provides for the following:

- postponement of the 2007 revaluation in England
- a new date for the revaluation to be made by order by the Secretary of State
- dates for any further revaluations in England thereafter to be made by such order
- repeal of the 10-year "rule" for future revaluations.

It may be noted that the 10-year cycle of revaluations for Wales is not affected by the Bill.

Valuation or Assessment for Council Tax

Aim

To examine the principles and practice of valuations for council tax

Objectives

- **to identify the roles and activities of participants in the assessment of dwellings for council tax**
- **to identify the definition of capital value and explore its meaning**
- **to briefly describe the procedure for council tax valuations**
- **to identify the basis of measurement for council tax purposes**
- **to identify the procedure for changing an entry in the valuation list**

Introduction

This chapter examines the basis of the way in which dwellings are valued for council tax purposes under section 22 of the Local Government Finance Act 1992. It begins with a consideration of such matters as roles and activities, measurement, the definition of "open market value" and dates.

However, whereas this chapter concentrates on the statutory basis of valuation, Chapter 16 considers the practical aspects of the valuation of a single property. The approaches to the original valuation list and the revaluations are dealt with in Chapter 14.

When the valuation matter involves a dispute as to value, the taxpayer is likely to have the expectation of a more precise result — this is the focus of the next chapter. It may be noted that related matters are covered in Chapters 17 and 18, namely:

- bodies involved in settling disputes
- appeals on disputed matters (but see p160)
- complaints.

<table>
<tr><td colspan="2">Box 15.1 Roles and activities the assessment for council tax</td></tr>
<tr>
<td valign="top">
Valuation Office Agency
• Listing officer (England and Wales)
• Assessor (Scotland)
Valuation and Land Agency Department
(of Northern Ireland)
• Valuation officer
</td>
<td valign="top">
• create and maintain the valuation list
• reference and record dwellings
• carry out revaluations
• make proposals to alter the list
• appeal before tribunals on behalf of the VOA
• maintain relationships with the public, billing authorities, listing officers of adjoining areas
</td>
</tr>
<tr>
<td valign="top">Referencer (in above offices)</td>
<td valign="top">• measures and records new buildings and buildings which have been altered</td>
</tr>
<tr>
<td valign="top">Valuer in the private sector</td>
<td valign="top">• may be appointed by the VOA to work on the revaluation
• advises taxpayer and others on value implications for council tax</td>
</tr>
<tr>
<td valign="top">Member of the valuation tribunal</td>
<td valign="top">• hears appeals on assessment matters from taxpayers</td>
</tr>
<tr>
<td valign="top">Regional director (VOA)</td>
<td valign="top">• if complainant not satisfied with office determination, takes the complaint for determination</td>
</tr>
<tr>
<td valign="top">Revenue adjudicator</td>
<td valign="top">• if the complainant not satisfied with regional director's determination, takes the complaint for determination</td>
</tr>
</table>

Roles and activities

Box 15.1 gives the principal participants in the assessment regime for council tax.

Valuation list alterations

Content of an entry following a proposal

The billing authorities' original valuation lists for England, Scotland and Wales contained details of the chargeable dwellings for each authority's respective area. Each entry was required to give the following information:

- a reference number given by the listing officer
- the address of the dwelling
- the valuation band of the dwelling
- for composite property a notation — "Comp".

Successful proposals in England and Wales result in alterations to the list which must be shown by:

- the date of an alteration
- a notation of the source of the alteration, ie valuation tribunal or the High Court.

Similarly, in Scotland the valuation list will show:

- the date of an alteration
- a notation of the source, ie the valuation appeal committee and the Court of Session.

Triggers to an alteration

Section 24(4) of the 1992 Act defines the circumstances following which the listing officer may alter the valuation list. Briefly, the defined circumstances are:

- there has been a "material increase" in value due to certain happenings and the event of a "relevant transaction"
- there has been a "material reduction" in the value due to certain happenings
- the property has become or ceased to be a composite property
- an increase or decrease of the domestic use of a dwelling of a composite property.

The circumstances are defined in section 24(10) of the 1992 Act. Again briefly, the terms mean:

- *material increase* — is one due to building, engineering or other operations to the dwelling
- *material reduction* — is one due to demolition of any part of the dwelling, any change in the physical state of the locality or adaptations for a physically disabled person
- *relevant transaction* — is the sale of a freehold, the grant of a lease of at least seven years' duration, and the sale of such a lease.

A material reduction in value due to part demolition is ignored if it is undertaken in the course works to the dwelling which are in progress or proposed. Box 15.5 below explores these issues in greater detail.

Any valuations needed to effect the alteration were originally to be dated at 1 April 1991 or as required subsequently (see section 21(2) of the 1992 Act).

Meaning of "domestic property"

Various enactments define "domestic" property including section 66 of the Local Government Finance Act 1988, ie the principal statute for non-domestic rating.

In brief, the legislation includes the following types of dwelling:

- property used wholly for living accommodation, ie sleeping, eating and associated purposes
- yard, garden, outhouse or other appurtenance belonging to or enjoyed with property used wholly as living accommodation
- private garage wholly or mainly for private motor vehicles (with the living space)
- private storage accommodation for articles of domestic use (with the living space)
- a caravan pitch or a mooring with caravan or boat respectively where the caravan or boat is an individual's sole or main residence (see section 7 of the 1992 Act)
- certain short-stay accommodation
- certain self-catering accommodation

- the residential part or parts of composite property
- dwellings associated with non-domestic property valued by formula
- dwellings on farms
- residential accommodation of public houses.

However, exclusions include property used for the business providing short stay accommodation, eg hotels (but see below).

Identification of property

It is, of course important that the listing officer correctly identifies property to be included in the valuation list. In most cases this will be obvious but problems may arise in respect of certain properties, such as the following:

- property laying across the borders of billing authorities (see pp72 and 178)
- caravans (see p182) and houseboats
- composite property (see p183)
- hotels, short stay accommodation and the like
- dwellings within a business valued by formula (see p183).

Hereditment

The council tax enactments incorporate (via the Local Government Finance Act 1988) the concept of the "hereditament" from the now repealed General Rate Act 1967. It means:

> ... property which is or may become liable to a rate, being a unit of such property which is, or would fall to be shown as a separate item in a valuation list.

Although other concepts like "rateable occupation" are not significant to council tax a brief look at the basis of the hereditament may be useful. Box 15.2 examines a selection of such matters. It draws on the work of Emeny and Wilks who explored the concept in considerable detail. It should be borne in mind that, on the one hand, the interpretation of the concept developed over very many years under the all embracing system of rating. On the other hand, council tax has a number of statutory features which replace aspects of the hereditament (see Appendix 6).

Basis of valuation for council tax

The valuations for the original valuation lists were undertaken by or supervised by the Valuation Office Agency as follows:

- the listing officer for each billing authority area in England and Wales
- the assessor for the area of the levying authority in Scotland (under the auspices of the VOA)
- for the original lists, valuers from the private sector who were contracted by the VOA (and supervised by each listing officer for the areas in which they worked).

Box 15.2 The concept of "hereditament" under the old rating regime

Should be within a parish	Originally a parish, but now the area of a billing authority — a property straddling a boundary is dealt with below
Should have a definable boundary	Example would be a house with a garden, say one miles away Case law: *University of Glasgow* v *Glasgow Assessor* (1952) SC 504 Held: several buildings within one fence were a single hereditament but other buildings of the university were single hereditaments *Peak* v *Burley Golf Club* (1960) CA53 R&IT Held: where part of a golf course could not be clearly identified the golf club was not rateable
Separated buildings have functional necessity to each other	Example: a house with road separating it from a garden and garage Case law: *Gilbert (VO)* v *Hickinbottom & Sons Ltd* [1956] QB 240 Held: that a bakery's two functionally related buildings which were separated by a highway constituted a single hereditament
There should be a single rate payer as occupier	The council tax legislation is very specific as to liable parties, eg owners who are not in occupation (in particular see Chapters 12 and 13)

The revaluations have or are being undertaken by the VOA.

The basis of valuation given below applies in the following circumstances:

- to assessments on revaluation undertaken by VOA personnel
- on appeal against the revaluation — following the coming into force of the valuation list there is a six months' period for appeals
- on subsequent proposals for a change, eg because the dwelling has been partly demolished (but see below).

For the last two kinds of valuations, private sector valuers may be instructed by owners, occupiers and others. They will seek to settle with the listing officer (see below).

Open market value as basis of valuation

Under the Domestic Property (Valuation) Regulations 1991 SI 1991 No 1934, the basis of valuation is open market value, ie capital value. (However, the charge to council tax is based on the capital value within a "band" (see Chapter 1). Generally, in the United Kingdom each dwelling is assessed to capital value for local taxation purposes by an official of the relevant government office.

In Great Britain the properties are entered on the valuation list by capital value band without an explicit value being shown or available. There is, therefore, not the pressure for taxpayers and others to appeal against an assessment that might otherwise occur, ie without banding. The rating system in Northern Ireland has no such banding — a discrete capital value being shown.

Value and valuation assumptions

The open market value of domestic property is defined in SI 1991 No 1934. It gives:

> ... the price at which the property might reasonably have been expected to realise if it had been sold in the open market by a willing vendor on (date)

This definition has a lineage based on the enactments for national capital taxes, eg estate duty (Finance Act 1894), capital gains tax (originally the Finance Act 1965) and capital transfer tax which was reformulated as inheritance tax (now the Inheritance Tax Act 1984).

Assumptions

The regulation 6 of the Council Tax (Situation and Valuation of Dwellings) Regulations 1992 SI 1992 No 550 provides the same definition and that the capital value (open market value) of a dwelling for council tax is assessed by the valuer on the assumptions given in Box 15.3.

It may be noted that the assumptions are not as straightforward as the above may suggest. In particular, regulations 6(3) and 6(3A) of SI 1993 No 550 may need to be observed.

Problems of analysis of comparables

In order to obtain an insight into the meaning of open market value one might consider case law on the definition in other contexts — capital taxes such as capital gains tax, inheritance tax and even estate duty (now defunct). However, there is a danger that an error may arise for one of the following reasons:

*   the assumptions, dates of transactions or other facts for the valuations in other statutory contexts are different from those in Box 15.2
*   the report of the case may not give sufficient information to make the appropriate analysis.

Nevertheless, the opinion of an experienced valuer may be preferred to subsequent bids and the price on a sale of the subject property within about six months of death. This follows the gist of the decision in *Commissioners of Inland Revenue* v *Clay* [1914] 3 KB 466 (the *Plymouth* case) — concerning increment value duty. However, the facts in this case concerned the purchase by the owners of an abutting property, ie as special purchasers, who wanted to extend their premises. The price probably reflected marriage value or development value or both. Both of these must be ignored for the purposes of an assessment for council tax (see p220 for *Site value rating*).

Composite property — "short cut valuation"

For a composite property, the basis of value of any dwelling within it is derived from the "relevant amount" of the whole property, ie its open market value on the assumptions in regulation 6 (taking

---

Box 15.3 Valuation assumptions for council tax

Vacant possession	The property is on the (notional) open market with vacant possession • this is the most common basis for houses and flats • other sale, eg with part possession will need adjustment
Tenure	The property is freehold, except in the case of a flat when it is leasehold for a term of 99 years at a nominal rent • say £1 (but this is not specified) • most house sales are freehold but not invariably so • flats are usually sold on lease — terms are usually different from the definition — they will require adjustment
Encumbrances	The property is being sold free of any rentcharge or other encumbrance • the agreed price may reflect the existence of an encumbrance and will therefore need adjustment • the sale price may not reflect the existence of an encumbrance
Physical aspects etc	The size, layout and character of the dwelling is as at the relevant date • but this is subject to conditions
Tone of the list	The physical state of the locality were the same as at the relevant date • but this is subject to conditions
Repair	The property is in a reasonable state of repair
Common parts (if any)	Where the occupier enjoys the use of common parts, they are in a reasonable state of repair • the purchaser contributes to their upkeep
Adaptations for disabled person	Adaptations for a person with disabilities which add to value are not included (see also Chapter 13)
Permanent restriction	Use would be permanently restricted to use as a dwelling • the actual sale price may reflect planning permission for a change of use (see *Development value*)
Development value	There is no development value, other than value attributable to permitted value (see Chapter16)

---

into account regulation 7). It seems that a short cut approach may sometimes be used; thus, in *Atkinson v Lord (LO)* [1997] RA 413, the Court of Appeal held that the valuer did not always need to find the relevant amount but merely the range of values within which it lies or a certain figure which it is above or below. (See p183 *et seq* for aspects of the valuation of composite property.)

Relevant date (or date of valuation)

Generally, any decrease in the value of a dwelling may give rise to an alteration of the valuation list by the listing officer. A decrease may happen in the following circumstances:

* when the property is demolished or partly demolished
* when the property is adapted for a person with disabilities
* when a deleterious change occurs in the locality.

Box 15.4 Antecedent valuation dates and relevant dates for council tax

Original valuation 1993
- 1 April 1991 (antecedent valuation date) (see reg 6 of SI 1992 No 550)
- 1 April 1993 (date for physical state and locality)

Revaluation 2007 (England)
- 1 April 2005 (antecedent valuation date) (see schedule 7 of the Local Government Act 2003) (but now postponed)

Revaluation 2005 (Wales)
- 1 April 2007 (date for physical state and locality)

Various events
- for alterations, the date of a subsequent sale (bearing in mind "tone of the list")
- for demolished buildings, the date of proposal
- for new buildings, the date of completion

However, there is a general embargo on proposals to alter the valuation list for increases in the value of an improved dwelling. This is an endeavour not to discourage improvements to dwellings. There are two exceptions, namely:

- when the property is sold by the owner who carried out the improvement
- when the next revaluation takes place.

More formally, the enactments provide for various dates in the valuation of dwellings for council tax, as follows:

- for revaluations two dates are important
- for other events, each will have a relevant date.

The various situations and their relevant or "antecedent" dates are summarised in Box 15.4.

Proposals

A valuation list may be altered by the listing officer as a result of

- information coming to the attention of the listing officer
- representations being made in the form of a "proposal" by the billing authority or other "interested person" and the representations are accepted or at least adopted in part by the listing officer as being well-founded.

When a change to property takes place, eg a part is demolished, it may have an impact on the value of the property, ie a reduction in value. The owner or occupier will normally seek to effect a change in the valuation list. If a proposal is successful, the listing officer will alter the entry for the property. Box 15.5 shows a schematic step by step procedure for a proposal.

Box 15.5 Step by step schematic procedure for a proposal

Step	Listing Officer (LO)	Interested person(s)/billing authority
1a	LO receives proposal and acknowledges it LO informs other IPs	Interested person (IP) makes a proposal to alter the list — to the LO
1b	LO considers the proposal invalid and serves an invalidity notice (but may withdraw it)	1b IP receives invalidity notice and accepts or appeals against it (it has not been withdrawn) to the valuation tribunal 1bb IP has four weeks within which to serve a notice of disagreement on the LO
2a	LO informs other IPs of receipt of valid proposal within six weeks	Other IPs receive notice of proposal — some may inform LO of wish to be a party to any proceedings
2b	LO receives notice of intention to be party to any proceedings	
3a	LO accepts the proposal and informs IPs by notice LO makes new entry in the valuation list	IPs receive notice and await change
3b	LO receives the withdrawal LO informs the other IPs (if they had expressed intention to be a party to the proceedings)	If the taxpayer, IP who made proposal withdraws it (If not now taxpayer, IP who made the proposal may withdraw it with the permission of present taxpayer)
4	LO informs billing authority of alteration to the Valuation list	Billing authority alters its copy of the valuation list
5	LO does not accept the proposal snd sets the appeal process in motion	

Completion notices

The creation of new dwellings by new-build or by conversions is likely to result in the issue of a completion notice by the billing authority in two kinds of circumstances, namely:

- when the billing authority considers that the dwelling will be completed within three months
- not having previously issued a completion notice, the billing authority does so because a new dwelling has become occupied.

Thus, section 17 of the 1992 Act and schedule 4A to the Local Government Finance Act 1988 provide the procedure for completion notices to apply to dwellings. For newly constructed dwellings or a major reconstruction, a completion notice is issued by the billing authority to the owner. The notice states the date on which the billing authority reasonably expect the building to be completed. The notice's date is the trigger for the billing of council tax which becomes payable three months later in the event of the building being unoccupied (the tax will be at the 50% discount). Of course, if the dwelling is occupied before the three months is expired the liability commences from the day of occupation. (Chapter 18 gives details of an appeal against a completion notice.)

Changes to property

Carrying out works of improvement to an existing dwelling or other events may result in the capital value of a dwelling changing. If the result is an increase in value, the entry in the valuation list should, in principle, result in a new entry in the valuation list but this is not usually immediate — the government, not wanting to discourage owners making improvements to property, only requires the entry to be changed as follows:

- when the property is sold (see regulation 4(1)(a)(i) of the Council Tax (Alteration of Lists and Appeals) Regulations 1993 SI 1993 No 290) (see p76 for *Buying and selling homes*)
- when the revaluation is undertaken on a national basis.

At these times a new assessment is made by the listing officer. If the new assessment is accepted by the owner, occupier or other interested parties, the banding will be altered accordingly. (However, there may be circumstances where the change could be challenged.) Box 15.6 sets out possible events and the possible effect on the property's council tax banding.

Generally, the state of repair and maintenance affects the condition of a property. For council tax the property will be assumed to be in a reasonable state of repair.

Inappropriate banding

On a revaluation, the new list is published by 31 December (say, September) to come into operation on the 1 April. An owner-occupier, landlord, tenant (as a taxpayer) or charging authority then has six months from 1 April to 30 November to appeal against the banding and seek a change. Appeals against the 1993 revaluation can, of course, no longer be made.

The appeals procedure is dealt with in Chapters 17 and 18. Briefly, for all appeals under the present system the listing officer may have initiated it by fixing a date with the clerk of the valuation tribunal prior to exhausting discussions with the proposer and any other parties involved. As a result the valuation tribunal may be convened only to find that cases are withdrawn because they were settled prior to the date of the hearing. If adopted, Appeals Direct will require that the valuation tribunal will only list cases which have exhausted resolution by negotiation between the parties.

Box 15.6 Works or other events: impact on council tax assessment

Use changed to non-domestic use	• listing officer should be informed • proposal to remove the property from the valuation list • may be entered on the rating list
Property becomes composite property	• a dwelling or other property becomes a composite property • will be entered on rating list (non-domestic part) and valuation list (domestic part)
Composite property changes	• proportion of floor space devoted to domestic use is altered
Demolition, part demolition or property no longer exists	• reduces the floor space • amend the valuation list
Improvements to property	• does not necessarily result in an increase • taken into account on a sale or on a revaluation
Works to meet the needs of person with disabilities	• applies to care homes as well as dwellings • see Chapter 16
Property is converted to two or more dwellings	• applies to a single dwelling being converted • applies to flats being created above shops and other non-domestic property • capital allowances may be available in the latter situation (see the Finance Act 2001)
Repairs and maintenance carried out	• generally, this should not affect the entry

Practical Aspects of Valuations

Aim

To introduce the assessment or valuation for council tax of domestic property — essentially houses and flats and other accommodation

Objectives

- **to outline the general features of an approach to assessing and valuing for council tax**
- **to identify sources of information relevant to council tax valuations**
- **to explain measurement of dwellings for council tax**
- **to identify problems associated with the analysis of transactions**
- **to consider the aspects of approaches to valuing different kinds of accommodation**

Introduction

It seems that there are at least two basic kinds of approach to assessment or valuation of houses, flats and other dwellings for council tax purposes, namely:

- the listing officer's appraisals for the original banding and subsequent bandings for the council tax valuation lists
- the valuations before or following a proposal which are undertaken by the listing officer or a valuer in the private sector (the latter acts on behalf of a client).

Other categories of property may require a special approach to arrive at an acceptable capital value, eg the living accommodation in a public house.

Essentially, this chapter examines the assessment or valuation, various kinds of property. It is given as an introduction to the information, measurement and approaches used by practitioners.

This chapter is best considered with the content of the following chapters:

- Chapter 14 — deals with the compilation and maintenance of the valuation list

- Chapter 15 — covers the basis of valuation and assessment
- Chapters 17 and 18 cover the resolution of disputes, ie appeals and complaints.

Assessment for the valuation list

When a valuation list is being prepared it is not required that a listing officer ascribes a specific value to a dwelling: the dwelling has to be allocated to a band (of capital values). The ascertainment for this purpose is that the estimate of value merely falls within the range of values for a particular band, eg the average band D. Of course, one might suppose that a more searching assessment will be made where, for instance:

- the property is unlike others in the locality
- the value of a property is expected to be near a change point on the scale of bands.

Antecedent valuation dates

The antecedent valuation dates are given in Box 15.4 (p170). For the original list and a revaluation, the assessments or "valuations" are undertaken and fixed with the antecedent date in the valuer's mind — to use an old expression, the tone of the list, ie the general juxtaposition of all values one to the other (albeit in bands). This applied to all dwellings existing at 1 April 1991 and seemingly identified new or altered buildings completed before the valuation list came into force on 1 April 1993. (However, it seems that in some areas the database was insufficiently robust to pick up all such cases.)

Valuation of single dwellings

Box 16.1 gives a notional procedure for the conduct of a valuation from the listing officer's (or representative's) perspective. For information, it will rely on the extensive and improved database held by the Valuation Office Agency (VOA).

Information for valuation

The Valuation Office Agency has a vast database of information on individual dwellings and other properties which are the subject of council tax, including composite property. As described in earlier chapters, eg Chapter 5, the information comes from various sources and has been and is further being developed, perhaps in part, as the National Spatial Address Infrastructure (NSAI) or ValueBill. This information may be used for valuation purposes by the listing officer and in certain circumstances by taxpayers or their representatives.

Powers of entry for valuations

Power of entry for valuation purposes is afforded to the valuation officer by section 26 of the 1992 Act. The officer must have a written authorisation and give at least three day's written notice of a visit to the dwelling. Entry is not permitted on weekends and bank holidays. For the new valuation list of

Box 16.1 Step by step schematic valuation procedure for a dwelling for council tax

Information search	• collect in-house information about dwelling • review information for sufficiency • draw up further information requirements
Information gathering	• if not obtained previously, dispatch forms for "particulars delivered" • request required information of likely holders • likely holders include the billing authority, the previous occupier or owner
Entry on the dwelling	• give notice of proposed entry for carrying out official valuation functions • gather information by entry on, eg condition survey and valuation data
Measurement	• follow the standard (see below) • deal with dwellings straddling boundaries
Property straddling borders	• having obtained measurements of the building and structure • determine the area containing the greater or greatest extent of the building etc within it • the dwelling is in that billing authority's area
Identify and analyse key properties	• identify dwellings or composite property which are similar to the subject property • gather in-house information • adjust the information and analyse the resultant data
Problems of analysis	• take care on the state of repair and condition • adjust for or eliminate "non-assumption" features or factors • find the unit of comparison • allow for dates of works or other "events" (see below) • for the above be aware of problems associated with analysis, particularly in the adjustments
Composite and other special property	• for composite property consider any special valuation requirements, eg the need for the "relevant amount" • these properties include caravans, boats, public houses, farms, dwellings with agricultural occupation conditions (see below)
Value the subject dwelling	• apply the unit to the subject dwelling's superficial area • if necessary, allow for "permitted development" (see below) • adjust the resultant value to obtain required result, eg for works for a person with disabilities (see below)
Model by computer (VOA)	• check the result by modelling the valuation results of the subject dwelling and other dwellings • "stand back" and review the result
Valuation list entry	• make the entry in the draft revaluation list or as an amendment to the existing list

1993, any private sector representative who undertook valuation work on behalf of the Valuation Office Agency was only able to enter the property outside the dwelling, ie there was no authority to go indoors.

Measurement for council tax

Measurements of property are needed for the following purposes:

- for council tax valuations on revaluation
- for council tax valuations following the creation of new buildings, the demolition of part of an existing dwelling, or the alteration or extension of an existing building
- for determining the location of a dwelling which lies across a border between two billing authority areas.

Different types of property are measured in different ways for council tax. Box 16.2 shows the measurements used for valuations of dwellings for council tax.

Box 16.2 Measurement for council tax — summary, based on the *RICS Code of Measuring Practice*

Measure	Tax	Type of property
Effective floor area	Council tax	Flats and maisonettes
Gross external area	Council tax	Houses and bungalows

Generally, purpose prepared plans will be required for construction and the laying out of grounds — many of these may be of use in measurement for council tax purposes. Users should be aware that metric measurement should be used for all applications for these purposes and for valuation for council tax. This is as a result of the Unit of Measurement Regulations 1995 SI 1995 No 1804.

Cross-border determinations

Plans produced by Ordnance Survey have many practical uses in the property industry, eg pinpointing location in property transactions and in planning applications. However, they will not normally be useful for council tax purposes, except perhaps for:

- locating a property
- making determinations of the location of a property which straddles the border between two billing authorities (see p72).

Permitted development

The valuation assumptions require development potential to be ignored other than that allowed as "permitted development". Such development is that in Parts 1 and 2 of the Town and Country (General Permitted Development) Order 1995 SI 1995 No 418, ie the GPDO. Box 16.3 gives a brief list of Classes A to H and of permitted development for a single dwelling house. Care should be taken to ensure that the rights are not restricted in some way, such as:

```
┌──────────────────────────────────────────────────────────────────────────────────────┐
│ Box 16.3 Single dwelling house: development covered by the GPDO                          │
│                                                                                          │
│ Part 1                                                                                   │
│ Class A      • enlargements, improvement or other alteration, including extensions alterations │
│ Class B      • roof additions or alteration which enlarge the house                      │
│ Class C      • roof alterations which do not alter the shape of the house                │
│ Class D      • porches or doors                                                          │
│ Class E      • building, enclosure swimming pool for incidental enjoyment                │
│ Class F      • provision of a hard surface, eg patio                                     │
│ Class G      • domestic heating oil storage container                                    │
│ Class H      • satellite antenna                                                         │
│                                                                                          │
│ Part 2                                                                                   │
│ Class A      • gate, fence, wall or other means of enclosure                             │
│ Class B      • means of access to certain highways                                       │
│ Class C      • painting the exterior, except of advertisements                           │
│                                                                                          │
│ Note: most of these works have restrictions affecting them, eg size, situation, height or volume │
└──────────────────────────────────────────────────────────────────────────────────────┘
```

- the dwelling is listed
- the dwelling is in a conservation area
- the dwelling is in a national park
- an article 4 direction affecting the GPDO is in force.

Adaptations for a person with a disability

An individual with a physical disability will often find that adaptations to their accommodation will make for their easier movement about the property. Typical changes may include:

- widening of doorways
- a lift between floors or a chair lift on stairs
- removal of thresholds
- for wheelchair access, gradual inclines or slopes between different levels
- special rooms for treatment, cooking, sleeping or ablutions
- handles and lifting apparatus in bathrooms and bedrooms.

In a valuation of a dwelling or care home for council tax purposes, any increase in value must be disregarded. (See Chapter 13 concerning the relief for adaptations for a person with disabilities, the so-called "band reduction relief".)

Box 16.4 Property transactions which may cause problems in an analysis

Dwelling on a "brown field" site	• property may have "hidden" problems, eg buried foundations
Dwelling with a garden which is known or thought to be contaminated	• price may reflect issues raised below in *Land with past contaminative uses*
Dwelling with development or hope value	• price for dwelling may reflect market's expectation that it will enjoy future development (hope value) • development value is considered in Chapter 19 (see Box 19.4)
Dwelling abutting a landlocked site	• price may reflect what the owner of the landlocked property paid to gain better access
Dwelling which is listed	• repairs etc tend to cost more
Other old dwellings in general	• (see Box 16.5) • the state of repair and condition needs careful attention • assume a reasonable state and condition • level of this may differ according to the age and location of the dwelling
Property where there is the prospect of marriage value	• price may have been paid by a sitting tenant — tends to be lower • price may not have been at arm's length, a low price was paid by a relative
Compulsory purchase of a property	• the price (compensation) may reflect the valuation rules and assumptions of the Land Compensation Act 1961 (as amended)

Analysis of prices of "problem" dwellings

Evidence of transactions may need to be analysed. In many cases this is straightforward. However, certain kinds of property are likely to present the valuer with practical problems when analysing prices for council tax. Properties with problems are shown in Box 16.4 (see Box 19.6 which, in relation to site value rating, covers some of the points given in Box 16.4 in greater detail.)

Land with past contaminative uses

Land which has been put to past contaminative uses may or may not be contaminated — the uses were well managed and no contamination occurred. However, a buyer can never be sure!

A dwelling which is contaminated or is likely to be contaminated raises a number of issues which may be reflected in the price, such as:

• the investigation and remediation will need to be paid for
• "the polluter pays" principle operates, but only where the polluter is known and is able to pay
• the effectiveness of any remediation may be uncertain
• any existing insurances, guarantees or warrantees may be or may become invalid
• after remediation it may still be difficult to obtain finance for works to the property
• the insurer may impose terms and conditions which are, as yet, unknown.

Landlocked sites

The price paid for a dwelling by the owner of an abutting landlocked site, ie land without adequate access to the highway may reflect some of the gain in its value to be gained by buying the dwelling — so giving better access. In valuing the back land, the original owner of the dwelling, ie the access land, may have argued for and received a proportion of the development value. Of course, if there are several prospective access routes, it may be possible to secure access more cheaply than would be possible where there is only one (see Box 19.6 which considers this point in the context of site vale rating).

Old property in general

In general, older property tends to have practical problems as shown in Box 16.5.

Box 16.5 Typical characteristics to be found in old dwellings

Lack of modern facilities	• installation and adaptation may be costly • "as is" state may be appropriate for the valuation rather than an assumed "modernisation"
Inadequate space	• may not provide for sufficient parking and other planning standards
Old structure and fabric	• may not allow easy modernisation
Out-of-date or old worn services and plant and machinery	• removal and renewal may be costly • obtaining parts may be costly or impossible • complete renewal may be required, eg electrical system
Asbestos	• requiring special treatment in removing it
Inadequate access and other facilities	• for persons with disabilities • works in accord with the Disability Discrimination Act 1995 and 2005 required
Structural weaknesses	• works required to strengthen the building
Vermin infestation	• removal may require costly specialist treatment
Listing	• listed building consent required for any demolition or works • labour tends to be in trades which are costly • materials and components tend to be costly • costs of any works tend to be higher

---

The valuer may be required to allow for matters like these in the analysis of a particular sale price. However, care is needed to ensure that the state and condition is not over allowed for in any adjustment — the valuer should have regard for the age of the property and its locality.

Box 16.6 Special considerations in valuation of certain kinds of dwelling

Crofts and farmhouses in Scotland	• where used with agricultural land • occupied by a farmer or farm workers • also, dwellings subject to an agricultural condition • assume that they are restricted to perpetual agricultural use as such
Dwellings in business valued on a formula basis, including: a lockkeeper's cottage	• within a business valued on a formula basis • although a composite property leave out the dwelling • value the dwelling separately as a dwelling
Licensed property with living accommodation	• the dwelling is usually part of a composite property • special valuation approach is adopted • licensed property, eg hotels and public houses, will often be composite property • in the billing authority areas of the VOA London Groups the London Specialist Rating Unit deals with licensed property
Dwellings as part of a farm	• usually involves consideration as part of a composite property • special valuation approach is adopted

---

Dates of valuation and effect of works or events

Where a new dwelling is completed after the coming into force of a new valuation list and has to be valued, the appropriate "antecedent valuation date" is used for the "tone of the list", ie the level of values are in the mind of the valuer. However, where for instance, a building has been altered — perhaps by partial demolition — the relevant date becomes important. Here, the valuer takes the property as it is at the relevant date and values it with reference to the values at the antecedent date.

Valuation points on particular property

Some properties need special consideration when being valued. They are listed in Box 16.6 with brief details of the issues to be considered. The section on *Composite property* below highlights the valuation issues in assessing that type of dwelling which is part of a larger property from which a business is conducted.

Caravan — sole or main residence

Under section 66(3) of the 1988 Act, a pitch and caravan used by the occupier as his or her sole or main residence, is a dwelling for council tax purposes. "Caravan" is defined by three statutes, namely:

• section 66(7) of the 1988 Act
• with reference to the Caravan Sites and Control of Development Act 1960
• amended by section 13 of the Caravan Sites Act 1968.

Where caravans are situated on pitches at a caravan or chalet park they may comprise:

- the proprietor's caravans to let (to holidaymakers)
- any caravans to let to holidaymakers which are owned by others
- caravans which are the sole or main residences of individuals.

The listing officer will assess the latter to a council tax band but give all of the proprietor's caravans a single non-domestic assessment, together with, perhaps, other lettable holiday caravans which are not owned by the proprietor (in accord with the special arrangements of regulation 3 of the Non-domestic Rating Regulations 1990 SI 1990 No 673, as amended). For certain sites and in certain circumstances SI 1990 No 673, as amended by the Non-domestic (Caravan Sites) Amendment Regulations 1991 SI 1991 No 141, will cover caravans which become non-domestic. (Class G exemption is considered on p146.)

Composite property

A composite property is one which is part business, ie non-domestic, and part domestic. In valuing the domestic element of a composite property two approaches are conceivable, namely:

- use direct comparison — analyse sales of nay similar domestic property in the locality and apply the outcome to the subject part property
- value the whole property — then apportion the value to the two parts, ie to the non-domestic and the domestic.

Reference may be made to the Council Tax (Situation and Valuation of Dwellings) Regulations 1992 SI 1992 No 550.

In *Salvation Army* v *Lane (LO)* (1994) CLVT (503029C/8) the listing officer used direct comparison and the Salvation Army's surveyor adopted the other method. The Central London Valuation Tribunal determined that the latter approach was appropriate with the result that the assessment was much lower, namely to band A rather than to band G.

Formula hereditaments

A non-domestic estate occupied by a utility or other organisation may be valued for business rating by formula — involving special calculations. The approach would embrace any composite property but the domestic element is excluded from the formula approach.

It is provided that any dwelling is left out of the hereditament and valued separately as a dwelling for council tax. In effect, the usual composite property approach is not used — the resultant value should reflect the dwelling as part of a larger unit.

Public houses

Several factors are considered when valuing living accommodation in a public house. In particular three pointers are used by the listing officer, namely:

- the open market values of suitable alternative accommodation in the locality

- the trading potential of the business of the public house relative to that of other houses
- the public house's accommodation and location, ie an exceptional location and accommodation will probably attract the market niche of buyers who may be expected to pay more than trade buyers.

The first appendix to the VOA's Practice Note 2 gives considerable detail of the methodology of the approach. Briefly, for non-exceptional properties the open market values of the suitable alternative accommodation are analysed into a range of bands. The result enables the identification of the "minimum band" and the "maximum band". For most public houses the band sought for the subject dwelling will be within the range. The trading potential is considered by reference to the "trading band". Public houses are traditionally valued by reference to trade. In the two years around the valuation date the officers of the VOA adopted the following methodology:

- analysed all sale transactions of such property in their entirety
- related the capital values to the 1990 range of rateable values
- produced a set of trading bands A to F where A was up to £200,000 capital value and F was over £750,000 or £800,000
- comparing the capital value of the subject property and the trading bands' range of capital values to obtain the adopted band for the subject property
- the result will not be below the minimum band or above the maximum band (except for exceptional case).

In the exceptional cases the open market vale for council tax will be found on the basis that the trading potential is not likely to be a significant factor from the hypothetical buyer's perspective.

Agricultural dwellings

This section briefly considers the valuation for council tax two kinds of property, namely:

- a dwelling or dwellings associated with a farm
- a dwelling which is subject to an agricultural planning condition.

Dwellings on farms are usually part of a composite property, but not exclusively (see p169). A detailed treatment of the valuation methodology is outside of the scope of this volume. However, the second appendix to the VOA's *Practice Note 2* deals with the methodology in some detail.

Box 16.7 identifies and briefly comments on the issues to be considered for the two kinds of dwelling mentioned above. In fact the issues are more complex than may be suggested in the box. The important point is to identify the extent of the composite property in terms of agricultural land and buildings together with one or more houses or cottages.

Basically, the assessment for each dwelling assumes that the whole is sold in one transaction and the hypothetical open market value attributable to each dwelling is found by apportioning that value.

The formula applied is as follows:

$$\frac{\text{Notional sale value of a house or cottage*}}{\text{Notional sale values of house, cottages and land*}} \times \text{Notional value of the whole farm}$$

where * indicates that the sale is by lotting.

Box 16.7 Agricultural dwellings and council tax

House or cottage
(actually sold as a non-agricultural
dwelling)
- treat as a normal rural dwelling
- subject to council tax without regard to this section
- buyer should ensure that of any planning condition concerning agricultural occupation is not operational

Farmhouse with adjacent agricultural
land and buildings
- a composite property
- notionally sold as a whole hereditament
- no lotting is assumed

Above, with farm one or more cottages
- cottage is occupied by farm worker
  — essential
- as before — not essential but better
  performance of duties
- as above — not essential not better
  performance

- all sold as part of a whole
- farmer liable for council tax
- part of composite property
- farmer liable for council tax
- part of composite property
- farm worker liable for council tax
- not part of the composite property

Farm house (or cottage) not with the land
which is disparately situated in the locality
- possibly treated as a separate property for council tax purposes
- not a hereditament, ie not a whole property
- a notional sale by lots may be assumed
- a notional sale as a whole is also assumed
- a ratio formula is applied to arrive at the values of each dwelling (see below)

Planning conditions and restrictions

Agricultural condition — single dwelling

Generally, this type of property is not a composite property. The local planning authority may have imposed an'"agricultural condition" on a dwelling which limits the occupier to one who is working in the agricultural industry. In valuing the property for council tax it is common to make a 10% to 30% deduction to take account of the condition. Whereas in Scotland the condition is taken as perpetual, elsewhere the prospect at the valuation date of the condition being lifted is taken into account. Another factor is the state of demand for such property.

Restrictions in law — exemption Class G

Some confusion exists concerning exempt property under Class G where occupation is prohibited by law (see p146).

Part 7

Dispute Resolution

Framework for the Resolution of Disputes

17

Aim

To show how the different kinds of dispute which arise in council tax matters are resolved

Objectives

- **to identify and distinguish different kinds of dispute**
- **to describe the different approaches to resolving disputes**
- **to apply the approaches to particular disputes**

Introduction

Although most cases on council tax matters go to the magistrate's court, the valuation tribunal or the benefit appeals tribunal, this chapter presents a range of approaches to dispute resolution which concern council tax. The kinds of dispute arise under council tax include:

- valuation
- challenge to the setting of the council tax
- the liability of a party
- mal-administration
- enforcement
- entitlement to council tax benefit.

This chapter reviews the organisation and structure for settling council tax disputes. Chapter 18 describes ways in which they may be settled. For the most part the disputes concern the taxpayer and some official body, eg the billing authority, but some may arise between the following:

- a landlord and a tenant
- an occupier and the listing officer
- an owner and a professional advisor.

Box 17.1 Roles and activities in dispute resolution

Claimant	• aggrieved person seeking a remedy to complaint
Defendant	• person complained about of alleged mis-doing
Judge	• a solicitor or barrister of requisite experience • appointed to be a judge by the appropriate authority • the court appoints a particular judge to hear a case • hears the evidence, reads the documents and hears the submissions of the parties or their lawyers • finds the facts and applies the law to reach the decision
Member of the valuation tribunal	• hears valuation appeals against proposals • must not be disqualified from being a councillor • must not be over 72 years of age • must not be employed by the tribunal • must not be a spouse of an employee
Arbitrator	• a lawyer or specialist professional in the field to be heard • appointed by the parties, a professional body's president • hears the case • determines judicially • not liable as arbitrator, unless acted in bad faith or resigned (when liability may arise)
Ombudsman	• appointed under statute or by an industry as knowledgeable of or experienced in the field • acts informally by correspondence • does not charge the complainant • establishes the whether complaint justified • award remedy, eg compensation
Adjudicator/Independent expert	• probably a professional in the field of the dispute • appointed by the parties • hears evidence from the parties and decides — based upon his knowledge and experience
Expert witness	• professional with knowledge and experience • must do the best possible for truth • may use books, specialist material and so on • gives opinion — has a duty to the court
Witness	• person with knowledge of facts and the parties, etc • including the listing officer and the recovery officer
Solicitor	• lawyer who handles the case for one of the parties • any barrister is instructed by the solicitor • may be an advocate solicitor able to appear in court
Barrister (see solicitor)	• lawyer who appears in court for one of the parties • presents the case, calling the witnesses • cross examines other party's witnesses • sums up on behalf of client

Box 17.2 Selection of disputed matters in council tax and the means of resolution

Construction
- completion notice
- discussion does not alter situation

- appeal to the billing authority (or assessor in Scotland)
- hence appeal to valuation tribunal or committee

Value of dwelling
- valuation of a dwelling
- LO's decision on proposal
- VT on a point of law

- listing officer's notice of alteration
- proposal to the listing officer (LO)
- appeal to the valuation tribunal (VT) on fact or point of law or both (VT is final on fact)
- appeal to the Court of Appeal, hence higher

Over-payment of tax
- repayment refused

- county court — civil debt action

Appeal against penalty
- England and Wales

- appeal or arbitration

Appeal against the granting of a liability order

- statement of case to the High Court

Appeal regarding an administrative or any discretionary decision of the billing authority that may cause injustice

- judicial review to High Court

Maladministration

- local government ombudsman (also known as the "Commissioner for Local Administration" in England)
- member of parliament
- regional director of HMR&C

Appeal against wrongful bailiff action

- county court

Appeal against a council tax benefit decision

- first appeal to the billing authority then if unresolved a further appeal to the benefits appeals service

Roles and activities

There are many participants in the different ways of resolving council tax disputes. Box 17.1 identifies the main roles and their activities.

Types of disputed matter

There are numerous matters which result in a dispute and a range of approaches are available to remedy them. In some instances the method of resolution is laid down by statute but sometimes a choice is available. Box 17.2 generally identifies each matter giving rise to dispute in council tax and council tax benefit. It also shows the possible approaches to resolving the disputes.

Criminal matters

The perspective of this volume is disputes on council tax and council tax benefits — essentially matters of a civil nature, eg proposals and appeals on the valuation of a dwelling. However, a few of these matters may involve theft or other offences. They are to be dealt with either by administrative sanction or penalty by the billing authority, or by criminal proceedings by the authority's in-house solicitor (see Chapter 10). However, criminal matters are generally matters for the Crown Prosecution Service (CPS) (in partnership with the police) and the criminal courts. The billing authority's counter-fraud team will liaise with the CPS and the police.

Proceeds of crime

There are some areas where those in the property industry have a duty under the law to handle appropriately suspected criminal matters. For instance, section 328 of the Proceeds of Crime Act 2002 (POCA) requires organisations to have in place arrangements for directors, managers and other staff to report situations where they suspect a person is using the proceeds of crime.

Other prosecutions (or otherwise)

It is unlikely that the POCA would be used in situations involving council tax but, in principle, it is available to the prosecuting authorities. However, other enactments have been used to prosecute in cases involving council tax. Theft and fraud are deal with in Chapter 10. Box 10.2 lists offences under several statutes. It may be noted, however, that each incident is dealt with within a hierarchy of administrative action or court action (see Chapter 10).

Approaches to dispute resolution

Box 17.3 gives the approaches commonly available in the property industry and includes brief description of each. These are dealt with in greater detail in Chapter 17 with particular reference to council tax and council tax benefit. However, compensation schemes and appeals are mentioned here.

Compensation schemes

Apart from the awards of compensation made by the ombudsman for loss due to injury or damage, there are other compensation schemes linked to local taxation. They include those linked to the complaints procedures of professional bodies, such as:

*   the Royal Institution of Chartered Surveyors
*   the Law Society
*   Institute of Revenues, Rating and Valuation.

Appeals

It may be noted that dissatisfaction with the decision of, for instance, a court, tribunal or ombudsman,

---

Box 17.3 Approaches to the resolution of disputes

Litigation	• a formal procedural approach of some gravitas in a court setting • many parties are represented but a litigant may find the other party is not represented • unrepresented litigants may add to the complications of the proceedings and the expense involved • loser may be required to bear the costs of both parties • judge and advocates are lawyers by profession • appeal to higher courts
Administrative action	• billing authority have administrative measures • caution • administrative penalties (a kind of fine) • tend to use them in less serious cases of fraud
Magistrate's Court	• hears many kind of enforcement or recovery cases • issues liability orders (see Chapter 18)
Valuation tribunal	• hears valuation appeals • determines completion notice disputes • determines on dwellings, eg sole or main residence, houses in multiple occupation
Arbitration	• determination likely to be by a specialist professional trained in arbitration • acts in a quasi-judicial manner, ie not using own knowledge and experience but settles on the basis of the evidence presented
Expert determination	• specialist professional may receive evidence • settles on the basis of own knowledge and experience • may be sued for negligence
Complaints procedure	• take complaint to the branch or firm • if dissatisfied, take to professional body or other independent determination • a particular scheme may provide for compensation, eg the Compensation Fund of the Law Society • does not cover negligence, fraud and similar issues
Ombudsmen	• there are four offices concerned with alleged maladministration (see Chapter18) • determination of complaints not settled by the billing authority, firm or company • relatively informal • power to award compensation

---

particularly on a point of law or on a matter of process, is a kind of "dispute" for which an appeal may be available, such as:

• from a valuation tribunal to the High Court
• from a billing authority's actions by way of judicial review to the High Court.

Box 17.4 Factors affecting the choice of the approach to resolving a dispute

Nature of the dispute	• the law provides the approach to settling some disputes, eg the courts or administrative measures • the contract between the parties may do likewise, eg a tenancy provides for the landlord to pay the council tax
Fees and costs	• courts may impose all or some costs against the person who loses the case • Civil Procedure Rules allow the court to penalise litigants who are "obstructive" • many costs are prescribed but, where not, debtors should ensure they are properly demanded
Speed of determination	• the speed of settlement may be essential
Insight and rigour of the determination	• rigorous legalistic approach tends to be taken by the courts • ombudsman's approach tends to be on a "fair and reasonable" basis • mediation is an approach where an expert guides the parties towards a settlement
Applicable law and jurisdiction	• the applicable choice of law and jurisdiction given in the contract is an important point to settle, probably by agreement
Relief or remedy sought	• (see below)
Degree of formality	• the range is high formality in the courts and relatively low formality in ombudsman's work • mediation, eg may be used to settle liability for council tax in a family dispute
Enforcement	• billing authorities have several ways of enforcing payment (see Chapter 8) • judgments of the courts may be enforced by execution against assets • contempt of court proceedings may be applied where a party acts against an order of the court • otherwise varying with the approach
Appeal	• appeal available in the courts (subject to permission perhaps) • appeal from an arbitrator award is limited, usually to procedural misconduct or an error of law • from the valuation tribunal to the High Court on a point of law, and hence to a higher court
Criminal courts	• fraud and corruption are dealt with by the criminal courts • billing authorities have administrative remedies and penalties — they are mainly used in the less serious fraud cases, for example

---

Box 17.5 Some of the reliefs and remedies available to disputants

High Court (see section 31 of the Supreme Court Act 1981)	• findings of a judicial review • order for mandamus, prohibition or certiorari • declaration or injunction • damages and award of costs may be made
Magistrate's Court (see Chapters 10 and 18)	• liability order and other remedies • attachment to earnings or allowances • committal to prison
Valuation tribunal (see Chapters 14, 15, 16 and 18)	• decision on appeal against assessment by the listing officer • disputes on various matters concerning buildings
Ombudsmen (see Chapter 18)	• findings on a complaint of alleged maladministration • redress on alleged mal-administration by billing authority or other public body
Professional bodies	• schemes for compensation

Factors affecting choice of approach

Generally, the risks inherent in any activity, project or procedure should be examined and dealt in ways which seek to avoid disputes. But that is an ideal world. However, in some instances of serious fraud, for instance, it is likely that the Department for Work and Pensions creates a deterrent effect in prosecutions (which are publicised) (see *Promotion of revenue services* in Chapter 5).

The parties in dispute may find that the choice of approach is determined by one of several means, eg the law, the contract between the parties. It is important, therefore, that the parties to any proposed relationship involving property should consider, among other things, the kinds of, or the ways in which disputes may arise. Approaches which could be used to resolve any dispute may then be more readily selected. The factors affecting the choice of approach are given in Box 7.4.

Relief or remedy sought

The purpose of any approach to dispute resolution is to satisfy at least one of the disputants with a relief or remedy. Box 17.5 sets out some of those that are available within the judicial and other systems.

Costs and fines

Many costs and fines are prescribed by enactment, eg fees which may be charged by a bailiff. However, the debtor, for instance, may need to confirm that an item in a demand for costs is properly allowable.

(This chapter has dealt with the principles for choosing a method of dispute resolution and some of the remedies — the following chapter covers some of the procedures of the magistrate's court, valuation tribunal and ombudsmen.)

Settlement of Particular Disputes

18

Aim

To show how particular disputes on council tax and council tax benefit are determined

Objectives

- **to consider magistrate's court procedures for the many disputes**
- **to consider disputes before the valuation tribunal**
- **to review how complaints are made and settled**

Introduction

Chapter 17 identified the main approaches to dealing with disputes. This chapter examines a selection of the approaches in some detail. A large number of potential disputes can arise under both council tax and council tax benefit. They are dealt with by the tribunals and courts and, in the case of complaints, by the various bodies concerned or one of the offices of the ombudsmen. Typical step by step illustrative procedures are given below. They are

- a recovery or enforcement dispute before a magistrate
- a valuation dispute before a valuation tribunal
- a complaint before an ombudsman
- a council tax benefit appeal.

Judicial reviews

In England and Wales a number of contentious matters are dealt with by judicial review by virtue of section 66(2)(a) to (e) of the 1992 Act, namely:

- specification of a class of "exempt dwelling" in an order
- determination of prescribed dwellings (see sections 8(2) and 12(1) of the 1992 Act)

Box 18.1 Step by step magistrate's court procedure to obtain a liability order

Step	Billing authority	Magistrate's court	Taxpayer
1	Books court time	Court time allocated	
2	Prepares "complaint list"	Receives complaint list	
3		Gives permission to issue summons(es) by signing the list	Receives summons
4	Serves summons(es) by ordinary post at least 14 days before hearing		

Interim period

5a	May receive request and negotiates to withdraw summons		May make request for withdrawal of summons and negotiate
5b	May negotiate but back any arrangement by continuing to apply for a liability order	Court encourages parties to make queries so as to: — keep costs down — speed up cases — settle cases	May find billing authority will accept promise of payment but insist on a liability order
5c	Not obliged, but may respond to queries and exchange information		Not obliged, but may respond to queries and exchange information

Day of hearing

6	Produces court list or revised complaints list, ie cases not settled or withdrawn		

Hearing

7	1 Officer under oath confirm requisites complied with and applies for liability orders	1 Deals with unopposed applications	1 Does not defend case — not usually in court
	2a Officer presents case for liability order	2 Hears defended cases	2a Defendant — presents defence — may question officer on authority's evidence
8		Decides cases — issues liability orders if the case is proved by council — awards costs as appropriate	
	Applies for costs in each case		

After court

	Billing authority: — sends letter demanding employment details — warns of the costs of bailiff action — if arrangement has been reached, retains liability order without further action, subject to payments being made as agreed — not obliged to make arrangement		Receives letter: — pays outstanding tax — pays according to an arrangement that may have been reached — if not, awaits further action

- original or substitute calculations of budget requirements (see sections 32 to 37, etc of the 1992 Act)
- a challenge to the setting of the council tax by the billing authority
- an original or substitute precept.

The court has power to grant relief and so quash the determination, calculation, setting or precept.

Liability orders before the magistrate's court

Liability orders relating to enforcement of collection are provided for by Part V of the Council Tax (Administration and Enforcement) Regulations 1992 SI 1992 No 613. Thus, if a reminder notice or a final notice has not been complied with, an enforcement dispute will be taken before the magistrates where the billing authority will apply for a summons and then seek a liability order (see regulation 23). Box 18.1 gives a schematic step by step procedure to the magistrate's decision. In practice several liability orders (perhaps thousands) will often be obtained with one application (see regulation 34). A detailed explanation is given below.

Booking court time

Before any application for a liability order can be considered, court time must be booked. The billing authority will either book an individual court sitting, or sometimes book ahead a sequence of courts sittings for a period of months.

Depending on the number of expected attenders, the authority will book anywhere between half an hour and a whole day of court time.

Complaint list

At least 14 days before the court date, the council, with the aid of its council tax data base, will produce a list of council tax accounts that have not been paid up to date after earlier notices. This list is known under different names in different areas, but most commonly is called a "complaint list". Permissions to issue summonses is granted by the court who sign the complaint list. The signatory is usually a magistrate or legal advisor with delegated powers.

Service of the summons(es)

The summonses are then served by ordinary post giving at least 14 days notice of the hearing (see regulation 35 of SI 1992 No 613). In practice the council and the court prefer to issue at least 15 days ahead if serving by first class post, and at least 16 days ahead if serving by second class post.

Withdrawing the summons

Between the issue of the summons and the court hearing, there may be dialogue between the debtor or their authorised representative and the billing authority. By this time the debtor will have lost their right to pay by instalments and the council, by way of the summons, is enforcing collecting of the full year's council tax. Representations may be made for the summons to be withdrawn and the right to pay by instalments reinstated. There is no statutory basis for this, and no obligation on the council to consider it unless there is error. However, policy varies from authority to authority and in some circumstances the summons may be withdrawn. Generally however, very exceptional circumstances will have to be shown, for the council to be able to justify this action.

Arrangement backed by a liability order

More commonly in this interim period a billing authority may receive a promise of periodic payments, but insist on still obtaining the liability order as a form of security so that in the event of further default, prompt enforcement action can be taken (such as bailiff or attachment of earnings (see Chapter 10)). The authority is not bound to make any such arrangement, nor is it bound by such an arrangement and can revoke it at any time.

Queries

There is no strict obligation on either party to disclose details of their case, but the court usually encourages the parties to make queries, and will also look to both parties to take reasonable steps to answer questions and keep down their costs.

Day of the hearing

Early on the day of the hearing, or sometimes the previous day, the billing authority will produce a "court list" or "revised complaint list" as it is sometimes known. This is a new list that shows all the cases that have still not either paid or been withdrawn. This list forms the basis of the bulk application for liability orders.

Hearing

Very few persons actually go before the magistrates to defend against an application for a council tax liability order. An unpublished study in Hampshire showed the figure to be around 1 in 500. This is due to four factors:

1   the billing authority will withdraw beforehand if the taxpayer raises a valid commonly recognised defence
2   council tax summonses usually being for less than £1500 are often paid in full before the hearing
3   there are very few valid defences
4   for those who do not wish to contest the matter, there is no reason to attend as there is no additional penalty for their absence.

Higher numbers attend the precincts of the court, but usually only to discuss matters with the billing authority's representatives in the hour prior the formal application, not to actually go before the magistrate.

Application for a liability order

An officer from the council, usually a senior revenue recovery officer, will confirm under oath that each of the necessary prerequisites have been carried out and make the application for liability orders and costs for each case that appears on the revised complaint list.

If there is a person or persons who wish to appear before the court, then sometimes the bulk application excluding those who wish to appear will be dealt with first, or sometimes the court will prefer to deal with the appearances first and then the bulk application.

When a person wishes to contest the application, the court will first hear from the council the specific evidence that relates to their case, and then invite them to respond. At hearings for liability order applications the parties rarely have questions for each other. Rather more often there is only representation to the magistrate. They may however put such questions, after each part of the evidence is given. If the court is satisfied with the billing authority's evidence and that there is no valid defence, then it must make a liability order.

Costs

The council will also apply for costs in each case. Typical costs for a liability order application are currently between £30 and £90. These are often made up of two parts, namely:

- costs incurred in the issue of the summons
- costs incurred in making the liability order application.

Usually costs follow the making of the order, though on occasion the magistrate will sometimes deny costs if it feels a degree of sympathy with the defendant. (Court fees are due to increase in 2006.)

Defences and "non-defences"

Common valid defences against the issue of a liability order include:

- bankruptcy proceedings or other form of insolvency proceedings have been instituted
- the level of council tax has not been set in accordance with the rules provided by the Local Government Finance Act 1992
- the billing authority has not applied the correct band as it appears in the valuation list
- the property on which the council is charging council tax does not appear in the valuation list
- the application is made more than six years after the council tax became due
- council tax has been paid
- there is a material defect in one or more of the prerequisite statutory notices
- instalments were not granted in accordance with regulations
- one or more of the statutory notices has not been properly served on the current or last known address
- the officer making the application has not been properly delegated with the authority to do so
- computer evidence is not supported by the appropriate certificate confirming the integrity of the source computer
- debtor is deceased
- the council has accepted part payment in full and final settlement

- there is relevant service under the Reserve and Auxiliary Forces (Protection of Civil Interests) Act 1951.

However, the following are not normally considered as valid defences:

- impecuniosity (insufficient money)
- name not spelt correctly
- any application, query, appeal or other matter relating to council tax benefit
- any appeal against the council tax banding
- an appeal against the listing officer's decision that the property is chargeable
- a challenge to the calculation of the specific bill
- alleged payment not substantiated by documentary evidence
- non-receipt of one or more of the statutory notices or summons
- actual or perceived shortcomings or failures in council services
- an amount for council tax was believed to be, or was actually paid to a landlord.

The reason why many of the above are not valid defences at a liability order application is because there are other more appropriate appeals processes. For example, a debtor must pursue unsatisfied benefit grievances at the benefit appeal service, and proposals or appeals regarding banding, designation as a chargeable domestic dwelling, or calculation of the bill must be made to the listing officer or valuation tribunal. Such a service or tribunal have the necessary specialist expertise to adjudicate on such matters.

Immediately after court

Under regulation 36 of SI 1992 No 613, the billing authority will send out letters to each person who has become subject to a liability order. Names of these letters vary but they usually have two effects. First, to demand details of the person's employment for the purpose of imposing attachment to earnings, and second, to inform the subject of statutory bailiff costs, which is a statutory prerequisite prior to instructing certificated bailiffs.

Box 18.2 gives a small selection of cases concerning issues that have arisen which relate to procedures for liability orders.

Committal before a magistrate's court

At times the billing authority will seek to have a defaulting taxpayer committed to prison. As indicated in Chapter 8, committal is often the last resort. A brief list of the steps in the committals process following an unsuccessful attempt at distress is given here:

- having been unsuccessful, the bailiff returns the liability order to the billing authority
- the billing authority sends a pre-committal warning letter (the letter is not required by enactment but is often a productive practical step none the less
- the billing authority books court time
- the billing authority prepares a summons and takes it to the court for signature
- if the defaulter does not attend court, the billing authority apply for an arrest warrant to enforce attendance

---

Box 18.2 Selection of cases concerning or relating to procedures for liability orders

Postal delivery — non-receipt of post	*R v Liverpool JJ, ex parte Greaves* [1979] RA 119 Held: it is not a defence that a summons was not received — that the summons had been posted was sufficient.
Council tax benefit	*R v Bristol City Magistrates Court & Bristol Council, ex parte Willman and Young* [1991] RA 292 Held: the existence of current council tax benefit queries, applications or appeals is not a defence to a liability order application
Refusal or remittance of case	*Hackney London Borough Council v Izdebski* [1988] RVR 114 Held: if there is no valid defence the court must make a liability order, ie it has no power to refuse or remit.
Exemption	*Tate (for Stockton on Tees District Council) v Yorkshire Egg Producers Ltd* [1981] RVR 253 Held: the magistrates may not determine an exemption
Valuation queries, proposals or appeals	*R v Thames Magistrate's Court, ex parte Christie* [1979] RA 231 *Hackney London Borough Council v Mott and Fairman* [1994] RA 381 *Shilto v Hinchcliffe* [1922] 1 All ER Rep 703 Held: as a defence, the magistrates cannot entertain challenges to the valuation band. They may not go behind the valuation list. They must give effect to whatever band has been determined. Thus, valuation queries, proposals or appeals are, therefore, not a defence to a liability order application (this is because there is a separate proposal and appeals process to the valuation office for such challenges)
Balance of probabilities	*R v Carr-Briant* [1943] KB 607, 612 Held: liability order matters, are to be decided on the balance of probabilities

---

- the warrant is usually granted by the magistrate, mostly with bail in the first instance, and without bail in the second instance of non-attendance
- on the day of the hearing the magistrate hears evidence from both parties
- the magistrate decides if there had been culpable neglect, wilful refusal or neither of these
- the magistrate decides whether to:
  1. commit to prison for up to 90 days
  2. commit to prison but suspend this on payment terms
  3. remit (write-off) part or all of the debt
  4. dismiss the case for the billing authority to make further attempts at enforcing collection or agreeing a repayment schedule.
- the decision of the court as to the existence of culpably neglect, wilful refusal, or neither, must be reached by conducting a sufficiently thorough means enquiry. This is established in a variety of cases including *R v Newcastle upon Tyne Justices, ex parte Devine* [1998] RA 97 and *R v Manchester City Magistrate's Court, ex parte Davies* [1989] RA 261

- Part I of the Criminal Justice Act 1982 applies to persons who face committal proceedings for unpaid council tax and are under the age of 21 years *R v Newcastle Justices, ex parte Ashley* [1993] RA 264
- attachment to earnings or deductions from income support must have been considered first, and shown to be impossible (such as in the case of the self- employed) or insufficient (such as in the case of the low earner, or person in receipt of non-attachable benefits such as incapacity benefit) *R v Highbury Corner Magistrate's Court, ex parte Uchendu* [1994] RA 51. More recently some courts are keen to see if the alternative of a charging order has been considered first
- in the last mentioned case it was held that the term of imprisonment imposed should be proportional to the degree of neglect of refusal.

Human rights and committal

Some committal proceedings have been the subject of claims of a disregard for human rights under the Human Rights Act 1998 and hence the European Convention on Human Rights. For instance, it has been held that defendants can be entitled to legal aid (see *Benham v United Kingdom* [1996] IHRL 36 (10 June 1996) in a judgment of the European Court of Human Rights) . Also in the same case it was held that the standard of proof needs to be "beyond reasonable doubt" because, while it is technically a civil matter, the potential outcome of up to 91 days in prison is both punitive and of the order of criminal matters.

Also, the registrar of the European Court of Human Rights in *Townsend v United Kingdom* (Application No 42039/98) (2005) The Times January 27 suggested that the parties try to settle their dispute amicably. They agreed the sum of £10,000 and costs for the infringement of the applicant's human rights. He had failed to pay tax according to the terms of a suspended prison sentence, but the summons to court to explain why was sent to the wrong address, and the court activated the committal order in his absence. He was therefore arrested and taken directly to prison. He claimed that his human rights under articles 5.1, 5.5, 6.1 and 6.3(c) of the European Convention on Human Rights were violated in that he did not have:

- the right to a fair hearing with a reasonable time
- the right to legal representation of one's choice
- the right to liberty and security.

Valuation tribunal appeals

Valuation tribunals derive their jurisdiction under the 1992 Act. Section 15(2) lists the enactments of jurisdiction, namely:

- section 16 of the 1992 Act — provides for appeals on certain administrative matters, eg that a dwelling is a chargeable dwelling
- the regulations under section 24(6) of the Act — concerning proposals and the accuracy of the valuation list
- schedule 3 to the 1992 Act, paragraph 3 — concerning certain penalties.

A valuation tribunal is convened by its clerk and usually comprises two or three lay members who consider a case in writing or at a hearing. The disputes they deal with include those on:

- appeals against completion notices
- the status of a dwelling, eg sole or main residence, house in multiple occupation
- appeals on valuations for council tax banding.

(Fuller accounts of the administration of valuation tribunals and the Valuation Tribunal Service is given below.)

Appeal against a completion notice

Initially, an appeal against a completion notice is made to the issuing billing authority. It is made in writing within 28 days of the date of issue of the notice by the billing authority. The appeal will usually be on the grounds that the building cannot be completed by the date specified in the notice.

The appeal should give details of the reasons why the date is considered erroneous, together with other information, eg to identify the appellant's status and the address of the property. The billing authority's decision should be received within two months. If the appeal is turned down or the billing authority's decision is not acceptable, an appeal may be made to the valuation tribunal within four weeks.

The appeal to the valuation tribunal may be taken in writing or a hearing will be held. The appellant may appear in person or be represented by a lawyer or surveyor.

Valuation appeals

Following a revaluation there is an opportunity on or before 30 November to make a proposal against the entry in the new valuation list by virtue of the Council Tax (Alteration of Lists and Appeals) Regulations 1993 SI 1993 No 290. Representations are made to the listing officer with the view to the list being altered. If the listing officer is minded to accept the representation (proposal), notification will be formally delivered to:

- the owner of the dwelling
- occuppier as a taxpayer
- the billing authority.

It will show the new assessment and the reason for the change.

In the first instance a council tax appeal against a listing officer's decision on a proposal is made to the valuation tribunal (which may confirm, vary, set aside, revoke or remit a decision or order) and hence to the High Court on a point of law. An appeal from a valuation tribunal decision must be made within four weeks. Appeals from the High Court go, with permission, to the Court of Appeal and hence the House of Lords.

On a national revaluation the right of appeal arises after the list has been published by the VOA and must have been made from 1 April and by 30 November (1 April 1993 for the original valuation list). Subsequently opportunities can arise in particular circumstances, such as:

Box 18.3 Reasons for changing the valuation list entry for a dwelling

Banding is inappropriate	• for instance, the property does not exist • the dwelling is in the area of another billing authority
Property is not domestic	• property is business property or otherwise not domestic, exclude from the valuation list
Alterations to property	• alteration eg partial demolition, may reduce its value below the lower threshold of the band • if so, make entry in a lower band • unless the property is sold, improvements between revaluations do not normally result in a new assessment between revaluation dates
Property has been demolished	• remove entry from the valuation list
Altered banding is not acceptable	• person aggrieved may seek to agree a new banding with the listing officer • appeal to the valuation tribunal
Planning permission has been implemented	• dwelling has been converted into two or more dwellings • dwelling has become a composite property • change of use to non-residential use • in Scotland, an agricultural condition has been imposed

---

- when there is a material reduction in the capital value
- when a property which has been improved by the owner is sold and the purchaser objects to the listing officer's new assessment.

Grounds for representation, ie a proposal, by an interested person or billing authority or reasons why the listing officer will make a change without being prompted, are shown in Box 18.3.

Procedure for appeal

The following principal steps illustrate an appeal to the valuation tribunal:

- following a notice of invalidity issued by the listing officer
- the proposer has four weeks to appeal — he or she serves a notice of disagreement on the listing officer
- the listing officer may withdraw the notice of invalidity or send the appeal to the clerk of the valuation tribunal stating the prescribed situation
- the valuation tribunal considers the appeal and makes its decision.

The procedures are more complicated than shown here but the gist of the appeal is given.

Valuation complaints (civil service)

Complaints about the civil servant's conduct in valuation matters may be referred to the local office in the first instance and then, if necessary dealt with in one of several ways. If dissatisfied with the local office a complaint is made to either:

- to the regional director of the Valuation Office Agency
- if dissatisfied again, to the Revenue Adjudicator

or

- to the local member of parliament.

Alternatively, a complaint may be made to the Parliamentary Ombudsman (though the local member of parliament (see below)).

Valuation tribunal administration

The 56 valuation tribunals in England and the four in Wales, replacing the valuation and community charge tribunals, are local bodies comprising three members drawn from a panel of local people. They hear appeals on the Valuation Office Agency's local taxation assessments of property, ie council tax (and business rates).

Where the valuation tribunal accepts the appeal it may direct the VOA to change the assessment and alter the list; otherwise it dismisses the appeal. A party to a hearing may appeal against the valuation tribunal's determination by going to the High Court on a point of law.

Administration and hearings

For a one-off appeal the clerk to a valuation tribunal arranges the hearing and during the sitting will advise the members on procedures and points of law, if necessary. Hearings are conducted in a relatively informal manner, usually with the ratepayer or taxpayer and the listing officer appearing in person.

The valuation tribunals have received revaluation appeals in Wales from 1 April 2005 and in England they will begin from 1 April 2007. The number of appeals should be at their highest sometime after the end of 2006.

Valuation Tribunal Service

The Valuation Tribunal Service (VTS) was established under the Local Government Act 2003 as an independent non-departmental public body (NDPB). From 1 April 2003, it provides administrative services and procedural advice to the valuation tribunals. Before the VTS was established, the staff of each valuation tribunal was, in effect, employed by the local chairman on terms and conditions which were not common to all valuation tribunals.

Valuation Tribunal Service for Wales

Draft regulations are about to be dealt with by the National Assembly for Wales. Essentially, it seems that the four valuation tribunals in Wales will constitute or have a separate service governed by a council but they should retain judicial independence.

Judicial independence

After consultation, the Office of the Deputy Prime Minister proposed in December 2004 to enhance the judicial independence of valuation tribunals. The Valuation Tribunal Board subsequently set up two committees, ie a Members' Judicial Committee (MJC) (of valuation tribunal members) and a Judicial Interface Committee (JIC) (of the valuation tribunals and the VTS).

Ombudsmen

A number of statutory offices of ombudsmen are related directly to council tax. Other ombudsmen deal with complaints which include or impinge on property matters indirectly concerning council tax. Thus, a complaint or dispute between a billing authority and a council taxpayer concerning maladministration may be resolved by one of the ombudsman for local government, namely:

- the Local Government Ombudsman (England)
- the Public Services Ombudsman (Wales)
- the Scottish Public Services Ombudsman.

For property in general, the ombudsmen who may deal with complaints involving parties in the private sector include:

- the Ombudsman for Estate Agents, to which an estate agent may join voluntarily
- the Legal Services Ombudsman.

(However, the Law Society has a compensation fund to redress loss caused by one of its members either acting dishonestly or failing to deal with a client's money appropriately.)

Complaints about civil servants, eg an officer of the Valuation Office Agency, may be dealt with by the Parliamentary Ombudsman. However, there are other routes for such a complaint (see p207).

General principles

Broadly, the principles under which each ombudsman works are common to each of their offices. Their services are free and informal. They usually conduct a complaint by correspondence, not requiring professional representation on behalf of the parties. It is an alternative to going to court and will not be available in cases where there has been a court hearing. Generally, any ombudsman's award is respected but is not necessarily enforceable by the complainant.

A typical step by step approach to taking a complaint to an ombudsman's office is shown in Box 18.4. Guidance on how to handle what is usually an informal process is available from each ombudsman.

Box 18.4 Step by step approach to taking a complaint to an ombudsman

Step 1	Approach the organisation	• explain the problem and seek redress from staff • if not solved, ask for official complaints procedure • if not solved, contact the ombudsman within set period of final response
Step 2	Approach to the ombudsman	• company's response is not acceptable or it has not responded in set period • receive advice on how to proceed
Step 3	Ombudsman sends complaint form	• complete the complaint form and send it in with relevant documents
Step 3	Follow ombudsman's initial approach	• ombudsman may seek to mediate informally • if not resolved, follow ombudsman advice
Step 4	Ombudsman appoints adjudicator	• adjudicator seeks further information from parties and others in carrying out a full investigation • adjudicator determines the outcome and informs the parties
Step 5	Ombudsman's award received	• if adjudicator decides for the complainant, up to set limits may be awarded

Council tax benefit decisions

Decisions by the billing authority on a claim for council tax benefit may be questioned and if a satisfactory answer is not obtained an appeal may be made to the benefit appeals service or the Department for Works and Pensions. A further appeal or an application for leave to appeal from a decision of the benefit appeals service may be made on a point of law to the Social Security Commissioners. Time limits are imposed on making these appeals.

It may be noted that the use of benefit review boards has been discontinued. The benefit appeals service was launched on 3 April 2000 to deal with most appeals relating to a variety of benefits including:

- housing benefit
- council tax benefit (from 2 July 2001)
- tax credit and child tax credit
- pensions credit
- child support
- social security.

Step by step benefit appeal

An applicant who believes that the authority is mistaken on his or her benefit may take the following steps:

- the applicant asks in writing for decision to be reconsidered by the authority (this request must be made within one month of the decision being notified). This period can be extended if the applicant had taken advice or initially sought a written explanation of the decision
- if necessary, the authority writes to applicant seeking more information
- the authority replies within 14 days with answer, including in the letter, details of how to appeal to the benefit appeal service
- if claimant believes that the council is still wrong, then they may apply to the appeals service under the Housing Benefit and Council Tax Benefit (Decisions and Appeals) Regulations 2001 SI 2001 No 1002 (this must be done in writing starting the grounds for appeal and exactly what it is that is believed to be wrong)
- the appeal is made via the council within one month of receiving the outcome of the council's reconsideration. (This period can be extended if the applicant has spend time taking independent advice, such as from a Citizens Advice Bureau)
- the authority will forward the claimant's letter giving his or her grounds for appeal to the benefit appeal service along with an account of the way the authority has calculated the claim
- the benefit appeal service will write to the claimant with a time and date for a hearing, and asking if they wish to attend in person or present their case in writing only
- after the hearing, the benefit appeal service will notify and the applicant and the authority of its decision
- if the person affected or the authority feels that the decision is wrong in law, they can seek leave to appeal to the Social Security Commissioners.

Appeals to the Social Security Commissioners

The Social Security Commissioners are lawyers appointed independently of the billing authority and the Department for Works and Pensions. They work to the Social Security Commissioners Procedure Regulations 1999 SI 1999 No 1495, as amended, which, *inter alia*, cover the appeals procedure on council tax benefit from 2 July 2001.

A form is available for the submission of an appeal, ie Appeal Form OSSC 1 within the correct time-limit. It should be sent to the Edinburgh or London for appeals in Scotland and for England and Wales respectively.

Applications must be received at these offices with the following:

- the decision notice
- the statement of reasons (the appeal must be made within one month of the claimant receiving this document)
- the notice of the ruling.

The application will be checked at the office and if appropriate it will be put before the Social Security Commissioners. They only deal with points of law: not questions of fact, nor findings or decisions; so it may be sent to another tribunal or even rejected. For instance, a late application will be rejected unless it is late because of exceptional circumstances or it was delayed for some reason which is considered special. Reasons for lateness must be given.

The appellant will be informed of the decision of the commissioners and it will normally be generally available on their website.

Part 8

Additions or Alternatives to Council Tax

Search for Additions or Alternatives

Aim

To examine the additions or alternatives to council tax

Objectives

- to examine each of the principal additional or alternative local taxes
- to consider the problems associated with each
- to review the repeatedly proposed service tax (income tax) for Scotland

Introduction

Many reviews of local taxation have been made. For instance, after consultation, the White Paper of 1981, *Alternatives to Domestic Rates* was published. It reviewed and rejected:

- income tax on a local basis
- payroll tax
- poll tax
- rates
- sales tax.

Now the government is again reviewing the future of local taxation and hence council tax. Hints of possible changes, such as replacement with a local income tax, have historically always been around. However, until a definitive policy statement is issued by the government and the policy is implemented, council tax is likely to remain without substantial change. In fact the indications are that the government currently intends to retain council tax and not reform it in the near future.

The Local Government Association and other bodies argue for a basket of taxes to fund local services. With this in mind some alternatives ways of extending the tax base are briefly considered below.

Community charge

In its Manifesto of 1987 the Conservative Party promoted a poll tax (called community charge) as an alternative to domestic rates. As a result, before the introduction of council tax, the Local Government Act 1988 introduced three kinds of community charge to replace domestic rates — they were:

- personal community charge — a "poll tax" on individuals
- standard community charge — a tax on second homes, paid by the owner
- collective community charge a tax on short stay accommodation, paid by the owner .

The community charge operated from 1 April 1989 in Scotland and 1 April 1990 in England and Wales. The new regime was, however, short lived and was itself succeeded by council tax. It was very unpopular, difficult to administer and, like the Peasants' Revolt of Wat Tyler's time, resulted in severe street violence in London.

Replacement of community charge

On 23 April 1991 the government's White Paper, *A New Tax for Local Government* proposed a new tax to replace the community charge. The Local Government Finance Act 1992 established council tax. It was a return to a dwelling or property based tax rather than a charge based on persons. However, council tax differs from the old domestic rating system, ie the one before the community charge. Council tax is based on bands of capital values whereas non-domestic rating was based upon rents (rateable values). By the time council tax was introduced, home ownership had greatly increased and although rent controls on dwellings were becoming markedly less, rental evidence was not so readily available as capital values — hence the change to capital assessments for council tax.

"Poll tax" element in council tax

It may be noted that council tax has a poll tax element in that a 25% discount is given when a dwelling has only one occupier. In effect, therefore, it could be argued that 50% of the of the council tax is attributed to the dwelling and the balance of 50 % applies to two or more adult occupiers (25% for the first occupier and 25% for the others — as the others leave the dwelling, leaving only one, a total charge of 75% is billed to the lone occupier. The tax is, thus, advantageous for households with three or more adults with earned or investment incomes. It might be argued that it is unfair (ie technically "regressive") on small households and more so perhaps on those households with low total incomes.

Conceivable extra payments

Billing authorities bill and collect council tax as an annual charge on owner-occupiers or tenants of residential property. It is payable on 1 April, or by instalments, and is subject to many exemptions and reliefs, eg the lone occupier relief. To spread the burden tax payments for, say, extra adults, lodgers and bed and breakfast guests are conceivable and have been mooted.

Cost of other bodies

Certain other bodies which are not billing authorities, eg police authorities, county councils and other bodies, make a budget for their services not funded from non-council tax sources. They then make a precept and in due course receive council tax from the billing authorities; the latter having added the precept when billing the tax payers.

Income tax (service tax)

Income tax has frequently been mooted as a local tax and attempts have been made in Scotland to replace council tax by service tax, a form of graduated income tax with exemption for those on low incomes. The latest attempt was provided under the Council Tax Abolition and Service Tax Introduction (Scotland) Bill (2004). However, it failed to pass into law.

The essential feature of the new tax was that it would have severed the direct link between local taxation and dwellings and other accommodation in Scotland. The burden of local taxation would have fallen on individuals who are resident in Scotland and are already caught under the United Kingdom income tax regime.

Box 19.1 shows the principal features of a local income tax (it is based partly on the failed proposal for service tax).

Box 19.1 Notional local income tax — possible features

Residency	• taxpayer would be a local resident in occupation • taxpayer with two or more dwellings will need be billed for all • exemption or relief may be given to non-main residences
Progressive quality	• income tax reflects the ability to pay and is generally fair between tax payers • if tax base is all sources, total income (earnings from employment, interest, dividends and rents) is the taxed
Regressive quality	• wealth which is not income producing, eg owner-occupied houses, objets d'art, antiques, development land is excluded • may distort the markets in such assets by making them more attractive relative to deposits, shares and the like • may cause a flight of capital to overseas
Ease of administration	• assessment, collection and enforcement systems are in place (HM Revenues & Customs) • evasion not likely to be any higher • difficult to link income tax to location of residence • taxpayer's change of residence will cause administrative difficulty • any differences of the local rate of income tax may cause difficulty

---

Effect of service tax

If the proposed service tax in Scotland had been adopted, the result would probably have been the removal of all RICS and IRRV members' involvement in valuation and property management matters on council tax. It is possible that some appeals would have been pursued but the Bill allowed for arrears of council tax to have been written off.

Similarly, the advent of service tax in Scotland would have had an affect on the organisation of local tax administrations. It is likely that collection of the tax would have fallen to the HMR&C. Generally, therefore, local authority involvement is likely to be lost in:

- billing and collection of council tax
- administration of council tax benefit
- making appeals against assessment
- recovery of unpaid council tax
- in the enforcement of the law.

Sales tax and other consumer charges

A sales tax is used in the USA but is not available for local taxation purposes in the UK. Value added tax (VAT) records could, conceivably, be used for a local taxation system. However, problems can be envisaged, including:

- the suitability of VAT records, eg across boundary problems where a supplier is in several locations
- the ability of a council to predict local sales each year
- the need for an averaging process by the central government, and hence the remoteness of the tax from local affairs.

The IRRV Committee of Inquiry has mooted the possibility of a sharing VAT revenue to local areas. Taxes, charges or fines to meet specific problems have been mooted and sometimes adopted. Provided the proceeds of such schemes are allocated to local authorities or to other public bodies, they could be regarded as payments for the costs of administration or law enforcement (see Box 19.2).

Consumer or user payments

Although the concept of payments for local goods and services provided by local authorities does not appear to be a comprehensive policy, it seems to be increasingly prevalent (see Box 19.3). Tourist taxes for overnight stays in an area are common abroad and could be introduced to spread the burden of council tax (see p214).

Traffic taxes

A part from any positive effects, the negative effects of vehicles are borne by local communities in terms of the following:

Box 19.2 Actual or potential ways to meet the cost of dealing with problems

Alcohol disturbance areas	• for a zone, suggested sales tax on businesses in the zone that sell alcoholic beverages
Congestion charges	• to raise funds for public transport • to raise funds for highway and amenity improvements
Highway safety charges	• to meet the costs of emergency and rescue services • to meet the cost of highway cleaning • to meet the cost of highway infrastructure repairs • to meet the cost of traffic calming
Parking space (on-site) levy	• to reduce traffic to workplaces and elsewhere
Planning charges	• to pay for planning services
Planning obligations or contributions	• to directly contribute to local infrastructure amenities and other improvements

Sales tax
- waste disposal — carrier bags
- removal from surfaces — graffiti by aerosol cans
- removal of gum from surfaces
* picking up litter, ie packaging, cans, bottles

- to encourage recycling or permanently useable substitutes
- to raise funds for the service

- to raise funds for the service
- to raise funds for the service

- loss of land and amenity when new roads are created or existing roads are widened
- wear and tear of road surfaces
- space use for parking
- clearing up after road accidents
- injury or death of local people and others
- deposit of substances and the release of fumes.

Compensation for road schemes

When a road is developed compensation is paid to residential owners/occupiers who have land taken from them or when they are affected by the use of the works but no land is taken. Generally, businesses and other organisations receive compensation when land is taken but receive nothing under Parts I to III of the Land Compensation Act 1973. As a result industrial and commercial workplaces, schools, hospitals and other local public assets may have to divert funds to mitigate, eg by the installation of noise insulation, the effects of a road scheme — without compensation.

Box 19.3 Actual or potential consumer or user payments for goods and services

Cultural facilities, eg theatre	• payment made for plays and other events
Education	• adult education is charged to the user • some evidence that some state schools are increasingly seeking "donations"
Leisure and sports facilities	• charged to the participant or user according to class or facility respectively
Events and services	• payment for policing services at football and other events
Roads and bridges	• tolls are charged to users
Shopping centres	• direct payment for policing services at large shopping centres
Waste	• garden waste is often removed for at the cost of a plastic bag • compost bins are sold to householders • removal of heavy item are often charged by the load or the item
Water	• standing charge and payment by volume (metered supply) • charge linked to council tax band but not normally to the billing authority's functions

Set-off

It may be noted that "set-off" applies to compensation on compulsory purchase. Thus increases in value due to the scheme are set-off against the compensation for land which is acquired compulsorily under the Land Compensation Act 1961.

Vehicle registration and road tax

Local vehicles are registered by a national body with an annual road tax. Conceivably a portion of the road tax comes back to the local community through national grants but no one knows and it is by no means certain. However, some known portion of the annual road tax raised in a community could, arguably, be allocated for local purposes.

Petroleum duties

Duties on petrol and other fuels are raised locally and taken into the national exchequer. Again, the Treasury could make some known specific allocation of fuel taxes to local authorities.

Road taxes

In India, the state governments impose taxes on vehicles about to enter their state. The levels and kinds of taxes vary from state to state. Generally, distinctions are made between cars, buses, and haulage vehicles and the number of passengers.

Road tolls and congestion charges

In a sense, petroleum duties are a tax on the use of roads but they are not high enough to discourage the use of vehicles by many people. Road tolls and congestion charges have a variety of uses, such as:

- road tolls may be used to pay the interest and repay the loan finance on public or private bridges, tunnels and the like
- congestion charges may be used to reduce congestion, ie reduce the use of vehicles
- both road tolls and congestion charges may be used to raise funds for public purposes
- both road tolls or congestion charges may be used to reward equity holdings in private toll schemes or to pay for any outsourced management for schemes.

However, the community near a toll facility may not be aware of local financial benefits, if any, accruing to their hierarchy of local authorities.

Payroll tax

There is already a form of payroll tax in the United Kingdom. National insurance contributions (NIC) are notionally dedicated to the National Health Service (NHS) and pensions. It may, therefore, be regarded as inviolate in this respect. However, apart from the local health services provided by the NHS, many local authority services are health-related and could, it might be argued, receive NIC.

However, if a payroll tax were to be introduced, it is more likely that it will be in addition to the NIC and have a wider remit than health. It would be relatively easy to administer and the tax-take would be stable and relatively difficult to avoid or evade. It would, however, be a tax on local business unless contributions were borne by work personnel (it might then be seen as a poll tax, payable by the employed).

Rating to rental value

Under the General Rate Act 1967 (and before) the rating of domestic property was based on the open market (net) rental value of property. For various reasons the approach was or became inadequate. These included:

- the artificial calculation of "statutory deductions" to arrive at net rent from gross rent
- the gradual national decline of the rental market in the private sector
- various approaches adopted by the government to control or regulate rents
- the rise of the public sector housing market at rents which were not fully open market rents (until the 1960s)
- the rise of owner occupation in the private sector
- the statutory right to buy provisions available to tenants of public sector landlords.

As a result the rental approach was abandoned when council tax was introduced — capital values were adopted and banded. (The IRRV Committee of Inquiry has suggested a return to discrete capital values (as prevail in Northern Ireland).)

Site value rating

In some countries, eg Australia and Denmark, "site value rating" is or has been used to raise annual funds for local services. While it has its adherents or promoters, it has never caught on as a means of local taxation in the UK. (Although development land tax (under the Development Land Tax Act 1976) might be seen as a form of site value taxation, the tax was paid on realised development value (as a once and for all payment). It was not designed to be a rating system, as such, for the funding of local government.)

Essentially, a site value tax would be charged each year according to the open market value of the land component of the property which might comprise both land and buildings — it may be at existing use value or it may include development value. The concept of development value is briefly described in the next few paragraphs. It could be used to tax, annually, enhanced land values due to say road, rail, underground rail and other public improvements.

Development value

The notion of "development value" is the key to understanding the ways in which some governments have introduced site value rating or have sought to tax or otherwise gather increases in value due to development, eg by planning obligation, planning gain or planning contribution.

Essentially, planning permission is seen to create development value (and any perceived prospect of obtaining planning permission "creates" hope value).

Obtaining planning permission for the development of a property usually, almost invariably, has a beneficial effect on its open market value. Box 19.4 shows the relationship of existing use value and open market value reflecting planning permission for development.

Box 19.4 Existing use value and development value

Example:
A property is a house with a large garden which it is expected could be sold in the open market for £200,000 as a house. However, with planning permission for several houses on the land, it could be expected to fetch £1,000,000 in the open market.

In technical terminology:	Open Market Value	=	£1,000,000
	Less Existing Use Value	=	£200,000
	Equals Development Value	=	£800,000

Numerous attempts to "tax" the capital, ie some or all of the £800,000 (or more accurately the development value) have been made since the end of World War II. However, site value rating is an annual impost. The following attempts, albeit unsuccessful in the long run are:

- development charge under the Town and Country Planning Act 1948 (1948 Act)
- existing use value compensation under the 1948 Act

- short term capital gains tax under the Finance Act 1962
- betterment levy under the Land Commission Act 1967
- development gains tax under the Finance Act 1974
- development land tax under the Development Land Tax Act 1976.

A somewhat hit and miss or arbitrary approach still exists in the following:

- planning gain or planning obligation — under various Town and Country Planning Acts
- conceivably, planning contribution under the Planning and Compulsory Purchase Act 2004.

However, the pre-budget statement (2005) indicates an intention to tax development value yet again.

Valuations

Valuations involving certain kinds of property with possible development potential are likely to present the valuer with particular problems (see Box 19.5).

Box 19.5 Valuation issues for property with possible problems

"Brown field" land	• tend to have foreseen and, possibly, unforeseen, hidden problems, eg contamination
Property which is known or thought to be contaminated	• the investigation and remediation will need to be paid for • "the polluter pays" principle operates but only where the polluter is known and is able to pay • the kind of development which would be permitted and the cost of remediation which would allow the said development • the effectiveness of any remediation may be uncertain • any existing insurances, guarantees or warrantees may be or may become invalid • after remediation it may still be difficult to obtain finance for the development • the insurer may impose terms and conditions which are, as yet, unknown
Property with hope value	• uncertainty is "paramount" • may be a lack of evidence
Landlocked sites (see Box 19.6)	• access may require acquisition of adjoining property • negotiation may lead to marriage value payment
Listed property with developable land	• presumption against the demolition • most development works including demolition require listed building consent as well as planning permission, if necessary • where works are permitted, the cost of the works tends to be relatively high
Old building capable of conversion	• may have unforeseen structural and other problems
Property with prospect of marriage value with an adjoining site	• a developer wants to extend site with adjoining land • marriage value may be negotiated in the price

---

Box 19.6 Valuation of landlocked development site — marriage value outlined

		£
Gross development value (of combined site)		G
Less Costs of development		(C)
Risk and profit		(P)
Land value (with access)	=	G – (C + P) = X
Less Value of site (without access)		(E)
Development value	=	X – E
Less Value of the site for access		(A)
Hence share for access dwelling @ say, 30%	=	0.3 (X – (E + A)) = Y
Open market value of access dwelling	=	Y + A

Notes

The share will depend on the availability of other possible access routes.

The owner of access land will want its existing value plus a share of the development value.

Listed buildings

The local planning authority usually requires the owner of a listed building to repair it to original styles and construction. Similarly, if allowed at all, alterations must comply with meticulous standards of conservation policy to ensure the changes are in keeping with the original building. Compliance with the standards tends to be relatively costly.

Old dwellings in general

In general, older properties tend to have problems. Box 19.7 shows the characteristics of old buildings in terms of structural and other defects.

The valuer may be required to allow for matters like these in the valuation. Of course, the nature of the valuation may enable them to be ignored completely. Thus, a different approach where it is

- a residual valuation assuming there is redevelopment after demolition and clearance of the site
- an insurance valuation on the assumption that the building would be replaced with a modern substitute building different approach.

Wealth tax

Wealth tax was mooted as a national tax in the 1970s but the idea was abandoned. Both Roman taxation and the Poor Relief Acts sought to catch the local wealth.

Thus, at about the time of the Poor Relief Act 1601, it appeared that local taxation embraced real and personal property, eg stock in trade. The *Earby* case of 1633 confirmed the tax base as real property and

---

Box 19.7 Typical characteristics to be found in old dwellings

Lack of modern facilities	• installation and adaptation may be costly
Inadequate space	• may not provide for sufficient parking and other planning standards
Old structure and fabric	• may not allow easy modernisation
Out-of-date or old worn services and plant and machinery	• removal and renewal may be costly • obtaining parts may be costly or impossible • complete renewal may be required, eg electrical system
Asbestos	• requiring special treatment in removing it
Inadequate access and other facilities	• for persons with disabilities • works in accord with the Disability Discrimination Acts 1995 and 2005 required
Structural weaknesses	• works required to strengthen the building
Vermin infestation	• removal may require costly specialist treatment
Listing	• listed building consent required for any demolition or works • labour tends to be in trades which are costly • materials and components tend to be costly

---

personal property within the parish but exclusions included wages, profits investment and furniture. Although stock in trade was part of the tax base it was not assessed consistently throughout the country and it was not included in assessments after 1840.

Conceiveably, "wealth" in an owner occupier's house could be used on death to pay unpaid accumulated council tax, eg as a land charge. It would be a life-time income relief — a form of annual equity release.

Wealth tax within council tax

Even today, council tax contains an element of wealth tax — albeit of a somewhat regressive nature. The tax is based on the notion that 50% of the capital value is attributed to the property and 50% to two persons residing in the dwelling (there is a 25% reduction when one of the two occupiers dies or leaves the property). The tax could, thus, be described as a diluted wealth tax which does not catch all the wealth, ie that held in the top bands, which could be available to it.

It could be said that a more comprehensive council tax, which not only included a person's wealth which is tied up in their home but also other assets, could be regarded as fairer than council tax (a partial wealth tax).

As noted above, both the Roman taxation system and the early form of rating under the Poor Relief Act 1601 covered tax bases which were wider than land and buildings alone. Not so council tax, which does not completely tax distinctions in the hierarchy of wealth tied up in homes, especially for the higher-value homes, eg homes worth several millions of pounds are taxed as those at about £360,000.

Thus, even if property values alone are taken, council tax aggregates all dwellings with a value over £360,000 as being equal in value by banding them in the Band H which, therefore ranges from £360,001 to at least, say, £45,000,000!

Stamp duty land tax

Stamp duty land tax is paid by the buyer of a dwelling on the price paid (on *ad valorum* basis), subject to exceptions. It is a kind of wealth tax but is not regular and is not available to the billing authority for local revenue purposes.

Easing regressive taxation

Council tax might be faulted on at least two counts, namely:

- it does not truly reflect ability to pay the tax out of income — a dwelling with say, five wages earners may be billed for the same amount of council tax as that for two retired persons in a similar house
- it does not reflect the capital wealth of the household in two ways —
  1. the capital values of dwellings range between band A and band G is roughly from £60,000 to £360,000 and that in band H ranges from the £360,000 to say £45,000,000 or more
  2. households with greatly different estates but similarly valued house are taxed on the family wealth tied up in the house not on the family's total wealth.

Of course, a wider form of council tax (by widening the base of wealth) would not be popular and may be relatively easy to avoid by "exporting" wealth. Also, it is likely to be less easy to assess and might fluctuate with the volatility which many assets exhibit.

European Union policies

Local taxation is a feature of all European Union (EU) countries and, as far as is known, there has been no suggestion that any harmonisation should be attempted. The EU policy of "four freedoms" and their concomitant state aid rules would apply to local taxation. If any local, regional or national government administration attempted to offer local taxation inducements which breached the EU free competition enactments, the European Commission might be expected to take action against the apparent infringement. However, it seems unlikely that domestic local taxes could substantially impact on major location decisions of international businesses — of course, local tax incentives might be part of a package of inappropriate state aid.

Future

The prospect of an immediate substantial change to local taxation seems to be receding as the writing of this volume nears its end (November 2005). The government has announced a kind of moratorium on change, eg the postponement of the revaluation in England. No doubt the report of the Lyons Inquiry as extended will be forthcoming at the end of 2006.

The future of local government seems to be in question and this may be reflected in the number of organisational changes and creations. These are perhaps generated by a more fundamental development of the information and communication technology — not unlike the later Victorian times when numerous public and voluntary bodies were created in endeavours to solve numerous problems urbanisation was generating as a result of the fundamentals of industrialisation.

In modern times new organisations are created under many guises, such as those generated for or coming from:

- arrangements under the private finance initiative (PFI)
- funding as projects by the National Lottery
- statutory arrangements to create the likes of the urban development corporations.

Partnerships

Some of the above are described as "partnership" arrangements. There are now partnership arrangements between bodies in all levels of government, the private sector, the voluntary sector and individuals. They are increasingly common.

In the context of local taxation they are engendered by the stakeholders seeking opportunities to utilise their own resources or garner the resources of others, eg Valuebill. The resources may be identified from the following:

- physical resources, such as property, furniture, office and other equipment and plant and machinery
- intellectual resources such as copyright material, patents and registered designs
- human intellectual, cultural and physical resources in the form of commercial specialist personnel and individuals with a range of more general qualifications, knowledge and skills, including volunteers
- financial resources.

Many bodies will no doubt see partnership arrangements in terms of commercial opportunities but others will adopt a perspective of joint enterprise to meet particular objectives which do not have a targeted commercial edge as such. Here the purpose is likely to be socio-economic development, cultural or social with perhaps either a public service dimension or a charitable intent.

The bottom line, however, is the utilisation of financial resources with greater efficacy and efficiency. In some ways the extension of range of objectives for charities, the seeming institutionalisation of the volunteering ethic and the changes to the roles of councillors may be examples of subtle changes in society, reflecting a need for a shift in the nature of local government — hence the extension to Lyons Inquiry. The items on charity objects and volunteering given below are briefly reviewed for their possible impact on the level of local council tax, ie the prospect that they will reduce the need for the public provision of services and hence tax funding.

Charity objects

Clause 2 of the Charity Bill extends the meaning of "charitable purpose" to include:

- the advancement of citizenship or community development

- the advancement of human rights, conflict resolution or reconciliation
- the relief of those in need by reason of youth, age ill-health, disability.

Some of these objects may touch upon and replace (in part or wholly) the work of local authorities which is funded to some extent by council tax.

Volunteering

The use of volunteers has been widespread for many years. Scores of local projects would not be sustainable without the input of volunteers' knowledge, skills and time. It is conceivable that the development of the voluntary sector could be extended to local projects and services. (For instance, in Lynchberg (Virginia, USA) local citizens maintain the street verges in a fashion which is similar to the voluntary disposal of targeted recyclable waste in the UK.) Volunteering England has been set up and has developed a strategy for volunteering for the community. It remains to be seen how the relationships between local volunteering organisations and local authorities will progress. From a local taxation perspective, local authority support may be seen as gearing the cost of local services, ie contributing to the cost of organisation for a greater return in the form of the volunteers' "sweat equity" (to borrow a term from the self-build fraternity).

New revaluation for England

The postponement of the revaluation and the removal of the cycle for revaluations (see p00) suggests reform of council tax has been put on hold until after the report of the Lyons Inquiry — perceived and actual unfairness in the system remains.

Although unlikely to give substantial revenues, other taxes, fines or charges are available and seem to be developing and others may, such as:

- congestion charges under the Transport Act 2000 (which are likely to be taken up by places outside London)
- fines for motorists caught by speed cameras (which seem to be increasing in number)
- workplace parking levy is available under the Transport Act 2000
- a proposed "development land tax" on planning permissions.

Conclusion

Local taxation has never been popular but there can be little doubt as to the virtue of the services that the council tax helps to finance. Every individual benefits from such services as environmental health, planning and housing to name a few — imagine society without environmental health standards or where a neighbour could develop a multi-storey apartments without adequate control. This volume has explored the continuous development of local taxation in the UK. Where continuous improvement, efficiency, effectiveness and fairness drive the delivery of local services, it is inevitable that local taxation will evolve to better meet the needs of the citizen, and to finance modern services — the report of the Lyons Inquiry is awaited.

Part 9

Appendices

Appendix 1

List of Boxes

Appendix 2
Table of Statutes

Table of Statutory Instruments

Appendix 3

Table of Cases

Appendix 4

Readings

Anderson, Lee et al (2005 updated) *Council tax manual*, Valuation Tribunal Service, (Electronic version)

Association of County Councils et al (1995) *Additional local taxes — Discussion Paper No 2: A local sales tax*, ACC Publications, London

Association of County Councils et al (1995) *Additional local taxes — Discussion Paper No 3: A local income tax*, ACC Publications, London

Audit Commission (2000) *On target — the practice of performance indicators*, Audit Commission (Electronic version)

Audit Commission (2005) *Consultation Paper — The framework for comprehensive performance assessment of district councils from 2006*, Audit Commission (Electronic version)

Audit Commission (2005) *Consultation Paper — The harder test — single tier and county councils framework for 2005*, Audit Commission (Electronic version)

Better Regulation Executive (2004) *Code of Practice on Consultation*, Cabinet Office, London (Electronic version)

Bond, Patrick & Brown, Peter (2002) *Rating valuation — principles into practice*, Estates Gazette, London

Cole, Vincent B, Parsons, Geoff(rey) B & Bell, Richard (1974) *Estate duty and property* 2nd ed, Estates Gazette, London

Departments: (Dates — various), HMSO, London

White Papers

(1988) *Modern local government: in touch with the people*

(1991) *A new tax for local government*

(2001) *Strong local leadership: quality local services*

(2004) *Transforming public services: complaints, redress and tribunals*

Department for Work and Pensions (2004) *Reducing fraud in the benefit system — achievements and ambitions* DWP, London (Electronic version)

Department for Work and Pensions (2005) *Touchbase: our newsletter for advisers, intermediaries and other professionals*, (Regular newsletter) DWP, Leeds

Eastman, Laurence S (1993) *Council tax*, Citizen's Advice Notes Trust, London

Emeny, Roger & Wilks, Hector M (1984) *Principles and practice of rating valuation*, 4th ed, Estates Gazette, London

Hinde, Thomas (Editor) (2002) *The Domesday Book — England's heritage then and now*, Greenwich Editions, Chrysalis Group plc, London

Hutton, Matthew (1999) *Tolley's tax planning for private residences*, 3rd ed, Tolley Publishing, Croydon, Surrey

Kneen, Pat & Travers, Tony (1994) *Implementing the council tax*, Joseph Rowntree Foundation, London

Lewsey, Christopher (Editor) (2004) *Encyclopedia of rating and local taxation*, (Update — May), Sweet & Maxwell, London

LPAC (2005) *LPAC newsletter* (Monthly — electronic version)

Miller, Kate (Managing Editor) (2005) *Insight magazine*, (Regular journal), Institute of Revenues, Rating and Valuation, London

Miller, Kate (Editor) (2005) *Valuer magazine*, (Regular journal), Institute of Revenues, Rating and Valuation, London

Miller, Kate (Editor) (2005) *Local taxation and revenues*, (Regular journal), Institute of Revenues, Rating and Valuation, London

Council tax and Valuation Tribunal Policy Team Leader (2005) *Council tax information letter*, (Monthly — electronic version) ODPM, London

ODPM (2005) (Draft statutory instrument — electronic version) *Home Information Pack 2006 SI 2006 No* (The number is not yet determined)

ODPM (2005) Consultation Paper *Inspection Reform: The future of local services inspection*, ODPM (Electronic version)

Office of Public Sector Information (2005) *Legislation*, OPSI (Electronic version)

Parsons, Geoffrey, Redding, Barry (1976) *Planning, development and the community land scheme*, Rating and Valuation Association (now IRRV), London

Parsons, Geoff (Editor) (2004) *The glossary of property terms*, 2nd ed, Estates Gazette, London

Parsons, Geoff (2005) *EG Property handbook*, Estates Gazette, London

Plimmer, Frances (1998) *Rating law and valuation*, Addison Wesley Longman Limited, Harlow, Essex

Restall, Mark (2005) *Volunteers and the law*, Volunteering England, London

Roots, Guy (General Editor) (2004) *Ryde on rating and the council tax*, 14th ed (Issue 34), Lexis Nexis UK (Butterworths), London

RICS (2001) *RICS Code of Measuring Practice: guide to valuers and surveyors*, 5th ed, Royal Institution of Chartered Surveyors, London

Slater, Ed (2005) (Updated) *Local taxation case law*, Institute of Revenues, Rating and Valuation, London

Slater, Ed (2005) (Updated) *Council tax law and practice; England and Wales*, Institute of Revenues, Rating and Valuation, London

Slater, Ed (2005) (Updated) *Local taxation case law*, Institute of Revenues, Rating and Valuation, London

Valuation Office Agency (2000) *Operational Instructions — Council Tax Manual — Practice Notes*, VAO, London (Electronic version) (see Box 5.3 for titles)

Ward, Martin (1994) *Council tax — handbook*, 2nd ed, Child Poverty Action Group, London

Ward, Martin, Zebedee, John (2003) *Guide to housing benefit and council tax benefit*, Shelter†, London

Appendix 5
List of Abbreviations

AC	Audit Commission
ACEA	Association of Civil Enforcement Agencies
AEO	attachment of earnings order
All ER	All England Reports
AME	annually managed expenditure
art	article
AV	average
BACS	bank automated clearing system
BC	borough council
BV	best value
BVPI	best value performance indicator
BVPP	best value performance plan
CAB	Citizens' Advice Bureau
CBA	Certified Bailiff Association
CC	county council
CIC	community interest company
CIPFA	Chartered Institute of Public Finance Accountants
CLVT	Central London Valuation Tribunal
com	commercial
Comp	composite (property)
CPA	comprehensive performance assessment
CPD	continuing professional development
CPR	Civil Procedure Rules
CPS	Crown Prosecution Service
CRS	compensation refund scheme
CT	council tax
CTAEO	council tax attachment of earnings order
CTB	council tax benefit

DC	district council
DCA	Department for Constitutional Affairs
DEL	departmental expenditure limit
DWP	Department for Work and Pensions
EAGGF	European Agriculture Guidance and Guarantee Fund
EG	Estates Gazette
EPA	enduring power of attorney
ERDF	European Regional Development Fund
EU	European Union
Four C's	Challenge-Compare-Consult-Compete
HB	housing benefit
HBRB	housing benefit review board
HBS	housing benefit security
HIP	home information pack
HM	Her Majesty's
HMO	house in multiple occupation
HMRC	Her Majesty's Revenue & Customs
IDeA	Improvement and Development Agency
IRRV	Institute of Revenues, Rating and Valuation
IS	income support
IUC	interview under caution
JIC	Judicial Interface Committee
JSA	jobseeker's allowance
KB	King's Bench
LA	local authority
LAIOG	Local Authority Investigation Officers Group
LBC	London borough council
LGFA	Local Government Finance Act
LGLR	Local Government Law Reports
LO	liability order; listing officer
MBC	metropolitan borough council
MJC	Members' Judicial Committee
NACABx	National Association of Citizens' Advice Bureaux
NDPB	non-departmental public body
NFI	National Fraud Initiative
NHS	National Health Service
NIC	national insurance contributions

NLPG	national land and property gazetteer
NoD	notice of determination
NNDR	national non-domestic rate
NSAI	national spatial address infrastructure
NVQ	national vocational qualification
ODPM	Office of the Deputy Prime Minister
PAF	postcode address file
Pt	Part
PDA	personal digital assistant
PI	performance indicator
PII	professional indemnity insurance
POCA	Proceeds of Crime Act
RA	Rating Appeals
reg	regulation
RICS	Royal Institution of Chartered Surveyors
RVA	Rating and Valuation Reporter
sch	schedule
SI	statutory instrument
SPARS	Sparsity Partnership for Authorities Delivering Rural Services
SPD	single person discount
TME	total managed expenditure
UN	United Nations
VAT	value added tax
VO	valuation officer
VOA	Valuation Office Agency
VT	valuation tribunal
VTS	Valuation Tribunal Service
VF	verification framework
YMCA	Young Men's Christian Association

Appendix 6

Concept of "Dwelling"

Statutes

In examining the concept of "chargeable dwelling" in Chapter 6, the statutory basis was considered briefly with references to particular provisions which were not addressed in detail. This appendix draws the various provisions together for a somewhat fuller treatment than could be justified in the chapter. Each section is referred to separately below and the details given briefly indicate the nature of the dwelling or other feature.

Local Government Finance Act 1992

- provides in a simple manner for "chargeable dwelling" and "exempt dwelling"
- it is in the complexity of section 3 of the 1992 Act where the meat of "dwelling" is found
- even section 3 uses provisions of other Acts
- finally the long list of "exempt dwellings" is found in the Council Tax (Exempt Dwellings) Order 1992 SI 1992 No 558 (see Chapter 13).

General Rate Act 1997

Section 115

- gives the concept of "unit" in the definition of "hereditament" ie
- "... property which is or may be liable to a rate, being a unit of such property which is, or would fall to be shown as a separate item in a valuation list"
- case law interprets the concept of hereditament (see Box 15.2).

Local Government Finance Act 1988

Section 66(1)(a)

- provides that a property is a dwelling if it is used wholly for the purpose of living accommodation, (note that if only part of a unit of property is used for living accommodation then that part will still be chargeable for council tax).

Section 66(1)(b) to(d)

- includes in a dwelling the likes of a yard, garden, outhouse, private garage and private storage.

Section 66(2)(a) and (b) excludes from dwelling commercially provided

- short-stay accommodation provided for short periods not used as a person's sole or main residence
- short- stay which is not self-contained self-catering.

Section 66(2A)(a) and (b)

- retains as a dwelling short-stay accommodation that in a person's sole or main residence whilst in residence and is not provided for more than six persons.

Section 66(2B)

- covers a commercially provided building or self-contained part as short-stay accommodation which is available for 140 days or more a year for short periods

Section 66(2D)

- covers a property which is a person's sole or main residence.

Section 66(2E)

- excludes timeshares covered by the Timeshare Act 1992.

Section 66(3) and (4)

- excludes any pitch with a caravan unless it is an person's sole or main residence and, similarly, a mooring with a boat.

Section 66(4A)

- last mentioned do not apply when they are appurtenant to a domestic property under subsection 1(a).

Section 66(5)

- property which is empty will be dwelling if it appears than its next use is likely to be a dwelling.

Section 66(9)
- the secretary of state may by order change the definition.

Local Government Finance Act 1992
Section 3(2)(a)

- resurrects the meaning of "unit" from section 115 of the 1967 Act for the purposes of council tax

Section 3(2)(b) and (c)

- provides that for a building to be a chargeable dwelling to council tax it must not fall in local or central "non-domestic" list or be exempt from non-domestic rating.

Section 3(3)(a) and (b)

- draws on Part III of the 1988 to include the domestic element of composite property in the council tax regime.

Section 3(4)

- excludes any yard, garden, outhouse, garage or private storage accommodation unless it is attached to a dwelling.

Section 3(6)

- empowers the Secretary of State to amend by order any definition of "dwelling".

Generally, the statutes provide that if a dwelling is self contained in so far as it has all its own usual domestic facilities, particularly for cooking and dining, then it is likely to be a separate dwelling for council tax This holds true even if the property:

- can only be accessed through another dwelling
- is prevented by planning regulations from separate sale or limited use.

Case law

It may be useful to bear in mind the substantial case law which had developed under the old rating system, both for domestic property and non-domestic property. Although council tax has a voluminous statutory basis, case law under rating need not be ignored in council tax. In particular, section 3 of the 1992 Act enables case law regarding what constitutes an individual chargeable dwelling for old style domestic rates to be carried forward to council tax. Nevertheless, council tax is developing its own case law on the concept of "dwelling".

Part 10

Indexes

Index 1

Organisations, Offices, Officials and Others

Index 2

Key Words